Claude Code: The Power User Guide

Terminal-Native AI Development

Ken Tannenbaum

2026

Finnoybu Press

Claude Code: The Power User Guide

2026 Edition

Terminal-Native AI Development

Published by Finnoybu Press

For permissions and inquiries: press@finnoybu.org

First Edition: 2026

About the Author

Ken Tannenbaum is a technologist, entrepreneur, and AI governance researcher. He is the founder of Finnoybu Press and the founder/steward of the AEGIS Initiative, an open framework for AI safety and accountability. Learn more at aegis-initiative.com.

When he's not building governance standards for artificial intelligence, he's helping everyday people understand and use it.

Dedication

For every developer who ever mass-deleted node_modules at 2 AM because they were convinced that was the problem. Spoiler: it wasn't.

And for everyone who saw a terminal-based AI and thought, *"Finally, someone gets it."* This book is for you.

The AI for Everyone Series

This book is part of the **AI for Everyone** series by Finnoybu Press, designed to help real people use AI tools effectively. Visit **press.finnoybu.org** for the latest titles and updates.

Collection One: Product Deep Dives

- *ChatGPT: The Definitive Guide (2026 Edition)*
- *ChatGPT: The Power User Guide (2026 Edition)*
- *Claude: The Definitive Guide (2026 Edition)*
- *Claude Code: The Power User Guide (2026 Edition)*
- *Gemini: The Complete Guide (2026 Edition)*
- *Gemini: The Advanced Guide (2026 Edition)*
- *Copilot: The Workplace AI Guide (2026 Edition)*
- *Copilot: The Advanced Workplace Guide (2026 Edition)*
- *Perplexity: The Research Playbook (2026 Edition)*
- *Perplexity: The Advanced Research Playbook (2026 Edition)*

Collection Two: Product Guides

- *NotebookLM: From Sources to Audio Overviews (2026 Edition)*
- *DeepSeek: Code, Math, and Reasoning (2026 Edition)*
- *Cursor: The AI Code Editor (2026 Edition)*
- *OpenClaw: Agentic AI for Everyday Work (2026 Edition)*

Collection Three: Cross-Platform Skills

- *AI Systems Playbook (2026 Edition)*
- *AI Prompt Engineering (2026 Edition)*
- *AI for Developers (2026 Edition)*
- *AI for Business (2026 Edition)*
- *AI for Students (2026 Edition)*
- *AI Governance for Practitioners (2026 Edition)*
- *Securing Agentic AI (2026 Edition)*

Table of Contents

Introduction 9

Part I: Getting Started 13

Chapter 1: What Is Claude Code? 15

Chapter 2: Installation and Setup 23

Chapter 3: Your First Session 33

Part II: Core Development Workflows 43

Chapter 4: Code Generation 45

Chapter 5: Code Review and Refactoring 55

Chapter 6: Debugging 65

Chapter 7: Documentation and Technical Writing 75

Part III: Project Management and Configuration 85

Chapter 8: The CLAUDE.md System 87

Chapter 9: Working with Large Codebases 99

Chapter 10: Multi-File Operations 111

Chapter 11: Testing Strategies 121

Chapter 12: Git and Version Control Workflows 133

Part IV: Extending Claude Code 145

Chapter 13: Hooks and Automation 147

Chapter 14: MCP Servers and External Tools 161

Chapter 15: The Claude Agent SDK 175

Chapter 16: IDE Integrations 189

Part V: Domain-Specific Development 201

Chapter 17: Frontend Development 203

Chapter 18: Backend Development 213

Chapter 19: DevOps and Infrastructure 223

Chapter 20: Working with Existing Codebases 235

Part VI: Operations and Collaboration 247

Chapter 21: Security and Safety 249

Chapter 22: Cost Management 259

Chapter 23: Team Workflows 269

Part VII: Reference 279

Chapter 24: The Evolving Landscape 281

Epilogue 289

Appendix A: Command Reference 291

Appendix B: CLAUDE.md Templates 299

Appendix C: MCP Server Directory 309

Appendix D: Troubleshooting 317

Appendix E: Glossary 325

Introduction

If you have used Claude on claude.ai, you know what it feels like to have a conversation with one of the most capable AI models available. You type a question, it responds. You paste in some code, it explains what is going on. It is useful, and for many tasks it is impressive. But there is a wall between that conversation and your development environment. You copy code from the browser, paste it into your editor, run it, find a problem, go back to the browser, explain the problem, get a new response, copy, paste, repeat. The AI is smart. The workflow is not.

Claude Code removes that wall. It is a command-line tool that brings Claude directly into your terminal, where it can read your files, write code, run your tests, execute shell commands, and interact with your entire development environment. It does not suggest changes for you to apply manually — it makes the changes itself, with your approval, and verifies that they work. The difference between chatting with an AI about code and having an AI agent operate inside your codebase is not incremental. It is a different way of working.

This book is the companion to *Claude: The Definitive Guide (2026 Edition)*, which covers Claude as a platform — its conversational capabilities, reasoning, and broad applications. This book is about what Claude can do with a terminal, a filesystem, and a real project to work on. Together with *ChatGPT: The Definitive Guide* and *ChatGPT: The Power User Guide*, this series covers the practical AI tools available to professionals today.

Who This Book Is For

This book is for developers who want to integrate AI into their development workflow — not as a novelty, not as a code snippet generator in a browser tab, but as a core part of how they build software. You should be comfortable with a terminal, familiar with git, and able to read code. You do not need to be an expert in any particular language or framework. The techniques in this book apply whether you write TypeScript, Python, Rust, Go, or anything else Claude Code supports.

If you have tried Claude Code once or twice and thought "there is clearly more here than I am getting out of it," this book is exactly what you need. If you have never used it at all, Chapter 2 will get you set up from scratch.

How This Book Is Organized

The book is structured in six parts that take you from installation through advanced mastery:

Part I: Getting Started (Chapters 1–3) covers what Claude Code is, how to install it, and your first hands-on session. By the end of Part I, you will have Claude Code running and a feel for the interaction model.

Part II: Core Development Workflows (Chapters 4–7) covers daily development: code generation, code review, debugging, and documentation.

Part III: Power User Techniques (Chapters 8–13) goes deeper into the CLAUDE.md system, large codebase strategies, multi-file operations, testing, Git workflows, and automation hooks.

Part IV: Extending Claude Code (Chapters 14–16) covers the Model Context Protocol (MCP), the Agent SDK, and IDE integrations.

Part V: Real-World Applications (Chapters 17–20) applies everything to frontend, backend, DevOps, and legacy codebase work.

Part VI: Professional Practice (Chapters 21–24) covers security, cost management, team workflows, and staying current as the tool evolves.

A Note on AI-Assisted Writing

This book was written with the assistance of AI tools, including Claude. The author directed the structure, content, and editorial decisions; AI assisted with drafting, research, and iteration. This is disclosed in the spirit of transparency and because it would be strange to write a book about AI-assisted development without using AI-assisted writing. The author reviewed and edited all content for accuracy, clarity, and voice.

A Note About Accuracy

Claude Code changes constantly. Anthropic ships updates regularly, sometimes weekly. Features get added, interfaces get redesigned, models get upgraded. This book is accurate as of May 2026, but specific details — menu locations, exact pricing, feature availability — may have shifted by the time you read this.

The principles, techniques, and strategies in this book have a much longer shelf life than the specific interface details. How to architect a CLAUDE.md, structure agentic workflows, and verify AI-generated code — those skills transfer across versions and even across different AI tools.

When something seems different from what's described here, check Anthropic's documentation at docs.claude.com for the latest information.

Let's Get Started

The terminal is waiting. Chapter 1 explains what Claude Code is, how it differs from the competition, and why the agentic approach to AI-assisted development represents a real shift in how software gets built. Let's get into it.

Part I: Getting Started

"First, solve the problem. Then, write the code."

— John Johnson

Chapter 1: What Is Claude Code?

You have probably used Claude before. Maybe you have chatted with it on claude.ai, asked it to summarize a document, or had it help you draft an email. That is Claude the conversational AI. Claude Code is something else entirely.

Claude Code is an agentic coding tool that lives in your terminal. It does not just answer questions about code — it reads your files, writes new ones, runs your tests, searches your codebase, executes shell commands, and builds features alongside you. If Claude on claude.ai is like texting a knowledgeable friend, Claude Code is like having a senior developer sitting next to you, hands on the keyboard, pair-programming in real time.

That distinction — between a chatbot and an agent — is the most important thing to understand before you go any further in this book. It shapes how you use Claude Code, what you can expect from it, and why it is a different tool from what you might have used before.

Claude Code vs. claude.ai vs. the Claude API

Anthropic offers Claude in several forms, and it is worth understanding where Claude Code fits.

claude.ai is the web-based conversational interface. You type a message, Claude responds. You can upload files, have multi-turn conversations, and use features like Projects and Artifacts. It is excellent for general-purpose tasks — writing, analysis, brainstorming, one-off coding questions. But it lives in a browser tab, isolated from your development environment. When you paste code into claude.ai, you are copying context manually. When Claude suggests a fix, you copy that fix and paste it into your editor. It works, but it is a manual, disconnected workflow.

The Claude API is the programmatic interface. You send HTTP requests with prompts and receive responses. It is what powers third-party tools, custom applications, and internal tooling. If you are building something that uses Claude under the hood, you are probably using the API. But the API by itself is just a language model endpoint — it does not know anything about your project, your files, or your terminal.

Claude Code sits on top of the API but adds an agentic runtime. When you launch Claude Code in your project directory, it understands the structure of your codebase. It can open files and read them. It can create new files and edit existing ones. It can run your test suite and read the output. It can search across thousands of files for a function definition. It can execute git commands, install dependencies, and run build scripts. It does all of this inside your development environment, not in a sandboxed playground or a browser tab.

Think of it this way: claude.ai knows what you tell it. Claude Code knows what your project tells it.

What Claude Code Actually Is

At its core, Claude Code is a command-line application you install via npm. You run `claude` in your terminal, and it starts an interactive session. From there, you describe what you want to accomplish, and Claude Code figures out how to do it — reading files to understand context, making changes, running commands to verify its work, and iterating until the task is done.

The word "agentic" gets thrown around a lot in AI circles, but here it has a specific meaning. Claude Code does not just generate text responses to your prompts. It takes *actions*. It has access to a set of tools — file reading, file writing, command execution, code search — and decides which tools to use, in what order, to accomplish your goal. You say "add input validation to the user registration endpoint," and Claude Code will:

1. Search your codebase to find the registration endpoint
2. Read the relevant files to understand the current implementation
3. Read your test files to understand your testing patterns
4. Write the validation logic
5. Update or create tests for the new validation
6. Run the tests to make sure everything passes

That is not autocomplete. That is not a chatbot. That is an agent executing a multi-step plan.

Under the hood, Claude Code is powered by Anthropic's Claude models. As of early 2026, that means Sonnet 4.6 as the default for most operations, Opus 4.7 for complex reasoning tasks, and Haiku 4.5 for lightweight operations. Sonnet 4.6

runs by default. For deeper reasoning, switch to Opus 4.7 with `/model opus` mid-session, the `--model` flag at startup, or the `ANTHROPIC_MODEL` environment variable. The `opusplan` alias is a useful middle ground — Opus for planning, Sonnet for execution. Claude Code falls back from Opus to Sonnet automatically only when you hit an Opus usage limit. Understanding these model tiers and when to use each one will matter when we discuss cost management in Chapter 22.

How It Differs from the Competition

You have options. The AI-assisted development space is crowded, and you should understand what makes Claude Code distinct from the alternatives.

GitHub Copilot pioneered AI code completion. Its greatest strength is inline autocomplete — you type a few characters, and it suggests the rest of the line or block. Copilot lives inside your editor, and its primary interaction model is suggestion and acceptance. It is excellent at what it does, but at its core it is a completion engine. It does not read your entire codebase to understand architecture. It does not run your tests. It does not execute multi-step plans. Copilot has added chat features and workspace agents over time, but its DNA is inline completion.

Cursor is a fork of VS Code with AI deeply integrated into the editor. It offers chat, inline editing, a "Composer" feature for multi-file changes, and codebase-aware context. Cursor is powerful and polished, but it is an editor — you have to use Cursor as your IDE to get its benefits. If you prefer a different editor, or if you work across multiple editors, Cursor's value proposition is tied to its environment.

Windsurf (by Codeium) takes a similar approach to Cursor — an AI-native editor with deep integration. Like Cursor, its strength is the tight coupling between AI and the editing experience. And like Cursor, that coupling is also its constraint.

ChatGPT's coding features have improved dramatically, especially with the Code Interpreter and Canvas features. But ChatGPT's coding support still operates primarily in a conversational context. You describe code, it generates code, you copy it over. It keeps getting better at understanding project context, but it does not live in your terminal and it does not have direct access to your filesystem.

Claude Code takes a different architectural approach from all of these. It is **editor-agnostic** — it runs in your terminal, so it works whether you use VS Code, Neovim, Emacs, IntelliJ, or any other editor. It is **agentic by default** — every interaction can involve reading files, writing files, and running commands, not just generating text. And it is **codebase-aware** — it can search and read your entire project to build context, rather than relying only on whatever files you happen to have open.

There is also the question of model quality. Claude Code is powered by Anthropic's Claude models, which are widely regarded as among the strongest available for code understanding and generation. The underlying model matters — the same agentic framework powered by a weaker model would produce weaker results. As of early 2026, Claude Opus 4.7 is one of the most capable coding models available from any provider, and that capability is what makes Claude Code's agentic approach work reliably rather than just theoretically.

That said, these tools are not mutually exclusive. Many developers use Claude Code alongside Copilot — Copilot for inline completions as they type, Claude Code for larger tasks that require planning and multi-file coordination. We will discuss hybrid workflows later in the book.

What Makes It Powerful

The power of Claude Code comes from the combination of four capabilities that, taken together, create something new in developer tooling.

Full codebase awareness. When you ask Claude Code about your project, it does not rely on whatever snippet you pasted into a chat window. It searches your codebase. It reads your package.json, your configuration files, your directory structure. It understands how your modules relate to each other. If you ask it to add a feature, it looks at how similar features are already implemented and follows those patterns.

Terminal access. Claude Code can run commands in your terminal — your build tools, your test runner, your linter, git, npm, pip, cargo, whatever your project uses. This means it can verify its own work. It does not just generate code and hope it compiles — it runs the compiler, reads the errors, fixes them, and tries again. This feedback loop is what turns code generation from "sometimes useful" to "reliably productive."

File read and write. Claude Code can read any file in your project and create or modify files directly. No copy-paste. No "here is the code, put it in your file at line 47." It makes the changes, and you review them. This sounds minor, but it eliminates a huge amount of friction from the AI-assisted development workflow.

Tool use and planning. Claude Code does not just execute a single action per prompt. It formulates plans, executes multiple steps, checks results, and adapts. If the first approach fails, it tries another. If it needs information it does not have, it goes and finds it. This agentic loop — plan, act, observe, revise — is what makes it feel less like a tool and more like a collaborator.

Integration and extensibility. Claude Code is not a closed box. It supports MCP (Model Context Protocol) for connecting to external tools and data sources — databases, APIs, documentation services, and more. It has a hooks system that lets you run custom commands before or after Claude Code takes actions, enabling automation workflows like auto-formatting on every file write or sending a Slack notification when a long task completes. And for developers who want to build their own tools, the Claude Agent SDK lets you create custom agents powered by the same underlying technology. We will cover all of these in Part IV of this book.

The Mental Model: A Senior Developer in Your Terminal

The most useful way to think about Claude Code is as a senior developer who has just joined your team. This developer is extremely fast at reading code, has broad knowledge of programming languages and frameworks, and is happy to do whatever task you throw at them. But — and this is crucial — they have never seen your codebase before.

This mental model gets you several things right:

You need to orient them. A new senior developer needs context. What does this project do? What are the conventions? Where do the important files live? With Claude Code, you provide this orientation partly through conversation ("this is a Django REST API for inventory management") and partly through CLAUDE.md files, which we will cover extensively in Chapter 8.

You should review their work. You would not merge a new team member's pull request without reviewing it, and you should not accept Claude Code's changes without reviewing them either. Claude Code is good — often remarkably good — but not infallible. It can misunderstand requirements, introduce subtle bugs, or make choices that do not align with your project's direction.

You delegate appropriately. You would not ask a new team member to redesign your entire architecture on their first day. Similarly, Claude Code works best when you give it well-scoped tasks. "Add input validation to the user registration endpoint" is a good task. "Rewrite the entire backend" is not.

They get better with context. The more a senior developer knows about your project, the better their contributions become. The same is true for Claude Code. A well-crafted CLAUDE.md, clear instructions, and good project structure compound into better results over time.

You communicate clearly. The quality of what you get from a colleague is directly tied to the quality of your request. "Make the app better" is a terrible task description for a human and an equally terrible prompt for Claude Code. "Add pagination to the /users endpoint, returning 20 results per page with cursor-based navigation" is specific, actionable, and likely to produce good results on the first try.

This mental model also helps you understand Claude Code's limitations. A senior developer who just joined your team might write perfectly functional code that does not match your team's style. They might not know about a tricky edge case that caused an outage last month. They might choose a library your team has deliberately avoided. Claude Code has the same blind spots — it knows what it can observe and what you tell it, and nothing more.

Who This Book Is For

This book assumes you know how to program. You do not need to be an expert — if you can write a function, read a stack trace, and use git, you have enough background. We will use examples in several languages throughout the book, but the concepts are language-agnostic.

You do not need prior experience with Claude Code, but this book will be most valuable if you have tried it once or twice. If you have installed it, run a few prompts, and thought "this is interesting, but I feel like I am only scratching the surface" — that is exactly where this book picks you up.

We are also not going to spend time convincing you that AI-assisted development is worth doing. If you are reading this, you have already decided to invest in learning these tools. Our job is to make that investment pay off as quickly and thoroughly as possible.

This book is written for developers across the stack. Whether you work primarily on frontend, backend, infrastructure, data pipelines, or some combination, Claude Code has workflows that apply. The examples throughout the book span multiple domains deliberately — we want you to see the breadth of what is possible.

What this book is *not* is a general guide to Claude. If you want to understand Claude's capabilities broadly — writing, analysis, reasoning, vision, and everything else — our companion title *Claude: The Definitive Guide (2026 Edition)* covers that territory. This book is laser-focused on Claude Code as a development tool.

What Is Ahead

In the next chapter, we will get Claude Code installed and configured on your machine. Then, in Chapter 3, we will walk through your first real session — not a toy example, but a genuine development task that shows you what Claude Code can do when it has access to a real codebase.

From there, we will work through the core workflows that make up daily development: code generation, refactoring, debugging, and documentation. Then we will move into advanced territory: CLAUDE.md mastery, large codebase strategies, the hooks system, MCP servers, and the Claude Agent SDK. By the end, you will not just know how to use Claude Code — you will know how to use it *well*, as a seamless part of your development workflow.

Let's get started.

Chapter 2: Installation and Setup

Getting Claude Code running on your machine is straightforward, but there are a few decisions to make along the way — how to authenticate, how to configure your environment, and whether to set up IDE integrations. This chapter walks through all of it, from a bare terminal to a configured setup ready for real work.

System Requirements

Claude Code runs on macOS, Linux, and Windows. Let's break down what you need for each platform.

macOS: You need Node.js 18 or later. If you are on a recent macOS version (Sonoma, Sequoia, or Tahoe), you are set. Install Node.js via the official installer at nodejs.org, or use a version manager like nvm or fnm. Most Mac developers already have Node.js installed — run `node --version` in your terminal to check.

Linux: Same requirement — Node.js 18 or later. Use your distribution's package manager or nvm. Claude Code works on all major distributions: Ubuntu, Fedora, Arch, Debian, and their derivatives. If you can run Node.js, you can run Claude Code.

Windows: As of 2026, Claude Code runs natively on Windows. Install Node.js 18 or later using the official installer at nodejs.org, and Claude Code will work directly in PowerShell, Windows Terminal, or Command Prompt — no additional layers required.

If you prefer, WSL (Windows Subsystem for Linux) remains a supported alternative. Some developers prefer WSL for its Linux-native shell environment. If you do not already have WSL set up, open PowerShell as Administrator and run:

```
wsl --install
```

This installs Ubuntu by default. Once WSL is running, install Node.js inside your WSL environment. Either approach — native Windows or WSL — works well. Choose whichever fits your existing workflow.

Beyond Node.js, you will want Git installed (Claude Code uses it heavily for version control operations) and a text editor you are comfortable with. Claude Code is editor-agnostic, so use whatever you prefer.

Installing Claude Code

With Node.js in place, installation is a single command:

```
npm install -g @anthropic-ai/claude-code
```

The `-g` flag installs it globally, which means the `claude` command will be available from any directory in your terminal. If you prefer not to install globally, you can use npx to run it without installation:

```
npx @anthropic-ai/claude-code
```

For regular use, the global install is more convenient. You will run `claude` dozens of times a day, and the npx approach adds a few seconds of startup overhead each time.

To verify the installation worked:

```
claude --version
```

You should see a version number. If you get a "command not found" error, make sure your npm global bin directory is in your PATH. On most systems, running `npm config get prefix` will show you where global packages are installed — the `bin` subdirectory of that path needs to be in your PATH environment variable.

Updating Claude Code

Claude Code is updated frequently — Anthropic ships improvements on a regular cadence. To update:

```
npm update -g @anthropic-ai/claude-code
```

Update every few weeks at minimum. New releases often include capability improvements, bug fixes, and support for new Claude model versions. Claude Code will occasionally nudge you to update if your version is significantly out of date.

Authentication

Claude Code needs access to Claude's language models, and there are two ways to authenticate: through a Claude Pro or Max subscription, or with an Anthropic API key.

Option 1: Claude Pro or Max Subscription

If you have a Claude Pro ($20/month) or Claude Max (Max 5x at $100/month or Max 20x at $200/month) subscription, you can authenticate Claude Code directly through your Anthropic account. The first time you run `claude`, it opens a browser window and walks you through an OAuth flow — sign in with your Anthropic account, grant permission, and you are set.

This is the simplest path for individual developers. Your Claude Code usage draws from your subscription's usage limits. With a Max subscription, you get significantly higher limits, which matters if you use Claude Code heavily throughout the day.

The trade-off: subscription access has usage caps. During periods of heavy use, you may hit rate limits, especially on the Pro tier. If you find yourself consistently bumping into limits, upgrading to Max or switching to API key authentication gives you more headroom.

Option 2: Anthropic API Key

For pay-as-you-go usage, you can authenticate with an API key. First, create an API key at console.anthropic.com. Then set it as an environment variable:

```
export ANTHROPIC_API_KEY=sk-ant-your-key-here
```

Add this to your shell profile (`.bashrc`, `.zshrc`, or equivalent) so it persists across terminal sessions.

API key authentication gives you usage-based billing with no caps — you pay per token, and you can use as much as you want. This is the better choice for heavy usage, team setups, or when you need predictable access without rate limits. We cover cost management in detail in Chapter 22, but for now, a typical day of active Claude Code use runs anywhere from $2 to $20 depending on how many tasks you run and which models are invoked.

Which Should You Choose?

If you are just getting started and want to explore, a Claude Pro subscription is the lowest-friction option. If you already have one for claude.ai, you can use the same subscription for Claude Code — no additional cost.

If you plan to use Claude Code as a core part of your daily workflow, a Max subscription or API key is the way to go. The Pro tier's limits will feel constraining once you are using Claude Code for real work throughout the day.

If your team or company is paying, an API key with usage-based billing makes cost tracking and budgeting straightforward.

First Run: The Onboarding Experience

With Claude Code installed and authentication configured, it is time to run it for the first time. Navigate to a project directory — any project you are currently working on — and type:

```
claude
```

On your first run, Claude Code will walk you through a brief onboarding. It will confirm your authentication, explain the permission model, and ask you to accept the terms of service. This takes about thirty seconds.

Once onboarding completes, you will see Claude Code's interactive prompt. It looks something like this:

```
Claude Code v1.x.x

  Welcome to Claude Code! I can help you with coding tasks in your project.

  /help for available commands

>
```

That > prompt is where you type your instructions. Try something simple to make sure everything is working:

```
> What files are in this project?
```

Claude Code scans your directory and gives you an overview. If it responds with a sensible description of your project's files, you are up and running.

Do not worry about learning the full interface right now — the next chapter covers it. For now, type `/exit` or press Ctrl+C to end the session.

Configuration: CLAUDE.md Files

Claude Code's most important configuration mechanism is the CLAUDE.md file. This is a Markdown file you place in your project's root directory (or in subdirectories, or in your home directory) that tells Claude Code about your project, your conventions, and your preferences.

You do not need a CLAUDE.md file to use Claude Code. It works fine without one. But a good CLAUDE.md dramatically improves the quality of Claude Code's output, and setting one up early is worth the investment.

To create a starter CLAUDE.md, you can use the init command:

```
claude /init
```

This generates a basic CLAUDE.md based on what Claude Code can observe about your project. It is a good starting point, but you will want to customize it. Here is what a CLAUDE.md might look like for a Node.js web application:

```
## Project Overview

This is a Node.js/Express REST API for inventory management.
Uses PostgreSQL for the database, Redis for caching.

## Conventions

- TypeScript strict mode
- Use async/await, never raw Promises or callbacks
- All API responses follow the { data, error, meta } envelope format
- Tests use Vitest with the convention: `*.test.ts` co-located with source files

## Architecture

- `/src/routes/` -- Express route handlers
- `/src/services/` -- Business logic
- `/src/models/` -- Database models (Drizzle ORM)
- `/src/middleware/` -- Express middleware
- `/src/utils/` -- Shared utilities

## Commands
```

```
- `npm run dev` -- Start development server
- `npm run test` -- Run test suite
- `npm run lint` -- Run ESLint
- `npm run build` -- Build for production
```

Chapter 8 goes deep on CLAUDE.md, including advanced techniques, templates for different project types, and strategies for evolving your CLAUDE.md over time. For now, just know it exists and that even a basic one helps.

Configuration Levels

CLAUDE.md files work at three levels:

User-level (`~/.claude/CLAUDE.md`): Your personal preferences that apply to every project. Things like "I prefer concise commit messages" or "always use American English in comments."

Project-level (`CLAUDE.md` in your project root): Project-specific conventions, architecture descriptions, and build commands. This is the most important one and the one you should create first.

Directory-level (`CLAUDE.md` in a subdirectory): Module-specific instructions. Useful in monorepos where different packages have different conventions.

Claude Code reads all applicable CLAUDE.md files and combines them, with more specific files taking precedence over more general ones.

Settings and Permissions

Beyond CLAUDE.md, Claude Code has a settings system that controls its behavior, particularly around permissions.

Claude Code's permission model is one of its most important features. By default, it asks for your approval before taking any action that modifies your system — writing files, running commands, deleting files. This is a safety measure, and a good one.

When Claude Code wants to perform an action, it shows you what it plans to do and asks for confirmation. You will see prompts like:

```
Claude wants to write to: src/utils/validation.ts
Allow? (y/n/always)
```

You have three options:

- **y** (yes): Allow this specific action.
- **n** (no): Deny this action.
- **always**: Allow this type of action for the rest of the session (or permanently, depending on the setting).

As you get comfortable with Claude Code, you will want to auto-approve certain types of actions. You can configure this in your settings. The settings file lives at `~/.claude/settings.json` and can be edited directly or managed through the Claude Code interface.

A reasonable starting configuration for an experienced developer might auto-approve file reads and code searches while still requiring approval for file writes and command execution. Chapter 21 covers fine-tuning your permission model when we cover security in depth.

IDE Integrations

Claude Code works great as a standalone terminal application, but if you spend most of your time in an IDE, the integrations are worth setting up.

VS Code Extension

Anthropic publishes an official Claude Code extension for VS Code. Install it from the VS Code marketplace by searching for "Claude Code" or by running:

```
code --install-extension anthropic.claude-code
```

The VS Code extension gives you Claude Code in a panel within your editor. You get the same capabilities as the terminal version — file reading, writing, command execution — integrated into your VS Code workflow. You can highlight code and send it to Claude Code, and Claude Code's file edits show up directly in your editor.

The extension uses the same authentication and configuration as the CLI version. If you have already set up the CLI, the extension will work immediately.

JetBrains Extension

If you use IntelliJ IDEA, PyCharm, WebStorm, or any other JetBrains IDE, there is an official Claude Code plugin available in the JetBrains Marketplace. Search for "Claude Code" in your IDE's plugin settings.

The JetBrains integration provides similar functionality to the VS Code extension: a panel for interacting with Claude Code, context from your editor, and direct file manipulation.

Terminal-in-IDE vs. Standalone

You do not have to use the IDE extensions to run Claude Code from within your IDE. Many developers simply open a terminal panel in their IDE and run `claude` there. This gives you the full CLI experience without extension-specific behavior.

The choice between the extension and the terminal comes down to preference. The extensions offer tighter integration — code highlighting context, inline diff views. The terminal gives you the raw CLI experience, which some developers prefer for its simplicity and consistency.

There is no wrong answer. Try both and see what feels right. Chapter 16 covers IDE integrations in full detail if you want to go deeper.

Verifying Your Setup

Before moving on, make sure everything is working. Run through this quick checklist:

1. Claude Code is installed:

```
claude --version
```

You should see a version number.

2. Authentication works:

```
claude
```

Claude Code should start without authentication errors. If you see an error about missing credentials, revisit the authentication section above.

3. Claude Code can read your project: Navigate to a project directory and start Claude Code. Ask it:

```
> Describe the structure of this project.
```

It should give you a sensible description of your project's files and directories.

4. Claude Code can run commands: Still in the session, ask:

```
> Run `git status` and tell me what you see.
```

Claude Code should ask for permission to run the command, execute it, and report the results.

5. (Optional) CLAUDE.md is in place: If you created a CLAUDE.md, ask Claude Code:

```
> What do you know about this project from the CLAUDE.md?
```

It should reference the information from your CLAUDE.md file.

If all of these work, your setup is complete. If any of them fail, here are the most common issues:

- **"command not found"**: Your npm global bin directory is not in your PATH.
- **Authentication errors**: Your API key is not set, or your subscription is not active.
- **Permission denied errors on Linux/macOS**: You may need to fix npm permissions. The npm docs have a guide on this — search for "npm resolving EACCES permissions errors."
- **WSL issues on Windows**: If you are using the WSL approach, make sure you are running Claude Code inside WSL, not in the Windows host shell.

Appendix D has a full troubleshooting guide for less common issues.

A Note on Multiple Machines

If you work across multiple machines — say, a laptop and a desktop, or a local machine and a cloud development environment — Claude Code's configuration is per-machine. Your `~/.claude/settings.json` and `~/.claude/CLAUDE.md` live on each machine independently.

Your project-level CLAUDE.md, however, lives in your project repository. If you commit it (and you should), it travels with your code. Your team can share the same project-level CLAUDE.md, ensuring everyone's Claude Code sessions have the same project context.

For personal settings you want to sync across machines, you will need to manage that yourself — a dotfiles repository, a symlink setup, or keeping them in sync manually. A minor inconvenience, but project-level configuration — the most important piece — syncs naturally through git.

What Comes Next

Your environment is ready. Claude Code is installed, authenticated, and configured. In the next chapter, we run a real session from start to finish. You will see how to interact with Claude Code, understand its tool use, work with the permission model, and get a feel for what a productive Claude Code workflow looks like.

Chapter 3: Your First Session

You have Claude Code installed, authenticated, and configured. Now it is time to use it. This chapter walks you through a complete session — not a contrived hello-world example, but a realistic workflow that shows how Claude Code thinks, acts, and collaborates.

By the end of this chapter, you will understand the conversation interface, the permission model, the core slash commands, and the rhythm of working with an agentic coding tool. You will also have a feel for when Claude Code shines and when you need to guide it.

Starting Claude Code

Open your terminal, navigate to a project directory, and type:

```
claude
```

That is it. Claude Code starts up, reads your project structure, loads any CLAUDE.md files it finds, and presents an interactive prompt.

A few things happen behind the scenes on startup. Claude Code scans your project's top-level files and directories to build an initial understanding of what it is working with. It reads your CLAUDE.md if one exists. It checks your settings for permission preferences. All of this takes a second or two.

You can also start Claude Code with an initial prompt, useful for scripting or when you know exactly what you want:

```
claude "add input validation to the signup endpoint"
```

This launches Claude Code and immediately begins the task. For now, start with interactive mode so you can see each step.

The Conversation Interface

Claude Code's interface is a conversation in your terminal. You type a message at the > prompt, Claude Code responds, and you go back and forth. Unlike a chatbot conversation, Claude Code's responses often include *actions* — reading files, writing code, running commands — interspersed with explanatory text.

A typical interaction looks something like this. You type:

```
> What does the authentication middleware do in this project?
```

Claude Code does not just guess based on the project name. It takes action. You will see something like:

```
I'll look at the authentication middleware in your project.

[Reading file: src/middleware/auth.ts]

[Searching for: authentication middleware references]

Based on what I found, your authentication middleware does the following:

1. Extracts the JWT token from the Authorization header
2. Verifies the token using the secret in your environment config
3. Decodes the user ID and attaches it to `req.user`
4. Returns a 401 response if the token is missing or invalid

The middleware is applied to all routes under `/api/` except
for `/api/auth/login` and `/api/auth/register`, which are
public endpoints.
```

Notice what happened. Claude Code did not just respond with pre-existing knowledge about authentication patterns. It *read your actual code* and described *your specific implementation*. This is the core difference between a chatbot and an agentic tool.

Your First Task: Exploring a Codebase

The best way to start any Claude Code session is to orient it in your project. Even with a CLAUDE.md, spending a minute or two helping Claude Code understand what you are working on pays dividends throughout the rest of the session.

Try these prompts as starting points:

```
> Give me an overview of this project's architecture.
```

Claude Code scans your directory structure, reads key files (package.json, go.mod, Cargo.toml, or whatever your project uses), and gives you a summary. This is useful not just for orienting Claude Code, but for verifying that its understanding matches your mental model. If it gets something wrong, correct it — that correction becomes part of the conversation context and improves everything that follows.

```
> What dependencies does this project use? Are any of them outdated?
```

This prompt triggers Claude Code to read your dependency files and potentially run a command to check for updates. It demonstrates two capabilities at once: file reading and command execution.

```
> Walk me through how a request flows from the API endpoint
  to the database in this project.
```

This is where Claude Code's ability to follow code across multiple files becomes valuable. It traces imports, reads function definitions, follows the call chain, and gives you a narrative description of the flow. For developers joining a new project, this is extraordinarily useful.

Understanding Claude Code's Capabilities

Here is what Claude Code can do during a session. Each is a tool Claude Code can invoke, and each can be combined with the others to accomplish complex tasks.

Reading Files

Claude Code can read any file in your project. It does this constantly — almost every task starts with reading relevant files to build context. You will see notifications like `[Reading file: path/to/file.ts]` as it works.

You do not need to tell Claude Code which files to read. When you describe a task, it figures out which files are relevant by searching your codebase, following imports, and reading configuration files. You can point it at specific files if you want:

```
> Look at src/services/payment.ts and explain the refund logic.
```

Writing Files

Claude Code can create new files and modify existing ones. When it writes a file, it shows you exactly what it plans to do before making the change. For modifications, it typically shows a diff — the lines it will add, remove, or change.

This is where the permission model comes in, covered in a moment. Claude Code will not modify your files without your approval (unless you have configured auto-approve for file writes).

Running Commands

Claude Code can execute shell commands in your terminal. This includes:

- Running your test suite (`npm test`, `pytest`, `cargo test`)
- Running linters and formatters (`eslint`, `prettier`, `black`)
- Git operations (`git status`, `git diff`, `git commit`)
- Package management (`npm install`, `pip install`)
- Build commands (`npm run build`, `make`, `go build`)
- Any other command-line tool available on your system

Command execution is one of Claude Code's most powerful features because it creates a feedback loop. Claude Code writes code, runs the tests, sees what fails, fixes the failures, and runs the tests again. No copy-paste workflow can match this iterate-until-it-works cycle.

Searching Code

Claude Code has built-in code search that works across your entire project. It can search by filename pattern, by content (like grep), or by semantic meaning. When you ask Claude Code to find something — a function, a pattern, a configuration value — it uses these search capabilities rather than guessing.

The Permission Model

The permission model is Claude Code's safety net. It ensures Claude Code never modifies your system without your knowledge and consent. Understanding how it works makes your sessions smoother.

Every action Claude Code takes falls into one of two categories: **read-only** and **write/execute**.

Read-only actions — reading files, searching code — are generally allowed by default. Claude Code does not ask permission before reading a file, because reading does not change anything. It needs to read freely to be useful.

Write and execute actions require approval. When Claude Code wants to write a file, create a file, delete a file, or run a command, it pauses and asks:

```
Claude wants to run: npm test
Allow? (y/n/always)
```

Your options:

- **y**: Allow this one time.
- **n**: Deny. Claude Code will adjust its approach.
- **always**: Allow this type of action going forward.

The "always" option is session-scoped by default — it lasts until you end the session. If you want to permanently auto-approve certain actions, you can configure that in your settings.

Finding Your Permission Comfort Level

When you are new to Claude Code, keep the defaults. Review every file write and every command execution. This builds trust in the tool and helps you understand what it is doing.

As you gain confidence, you will want to relax permissions for common, safe operations. A reasonable progression:

Week one: Approve everything manually. Watch what Claude Code does. Get a feel for its judgment.

Week two: Auto-approve test execution and linting commands. These are read-only in spirit — they check your code but do not change it.

Week three: Auto-approve file writes in your source directories. You are still reviewing changes, but you are trusting Claude Code to make them directly rather than asking permission each time.

Ongoing: Periodically reassess. If you find yourself always approving a certain type of action, auto-approve it. If you notice Claude Code making mistakes with a certain type of action, tighten the permission.

There is no universally correct permission configuration. It depends on your risk tolerance, your project, and how much you trust Claude Code with your specific codebase. What matters is that the system gives you control.

Slash Commands

Claude Code has a set of slash commands — special commands prefixed with / that control the tool itself rather than give it a task. Here are the ones you will use most often.

/help

```
> /help
```

Displays a list of available commands and a brief description of each. This is your reference card. When you forget a command, start here.

/clear

```
> /clear
```

Clears the conversation history and starts fresh. Claude Code retains your CLAUDE.md context but forgets everything from the current conversation. Use this when you switch to an unrelated task and do not want the previous conversation polluting Claude Code's context.

/compact

```
> /compact
```

This is one of Claude Code's most important commands, and one new users often overlook. It compresses the current conversation, keeping the essential information but reducing the token count. This is critical for long sessions covering multiple tasks.

Every message in your conversation costs tokens — both for the input (what Claude Code reads) and the output (what it generates). As your conversation grows, each new message becomes more expensive because Claude Code rereads the entire conversation for context. The /compact command solves this by summarizing the conversation so far into a condensed form.

A good rule of thumb: if you have been working for more than 20–30 minutes on a continuous session, run /compact. Your token bill will thank you.

/cost

```
> /cost
```

Shows you how much the current session has cost in terms of token usage. If you are on an API key, this translates directly to dollars. If you are on a subscription, it shows you token consumption relative to your limits.

Get in the habit of checking /cost periodically. It builds intuition for what kinds of tasks are expensive (large codebase analysis, multi-file refactoring) vs. cheap (quick questions, single-file edits).

/status

```
> /status
```

Shows the current state of your Claude Code session: which model is active, your authentication method, loaded configuration, and other diagnostic information. Useful for debugging when something is not working as expected.

/init

```
> /init
```

Generates a CLAUDE.md file for your project based on what Claude Code can observe. Best used at the start of a new project or when setting up Claude Code in an existing project for the first time. We covered this briefly in Chapter 2 and will go deep in Chapter 8.

A Complete Worked Example

Let's walk through a realistic task from start to finish. You have a Node.js Express API and want to add rate limiting to your public endpoints.

You start Claude Code in your project directory:

```
claude
```

You give it the task:

```
> Add rate limiting to the public API endpoints. Use a sliding
  window approach, 100 requests per minute per IP address.
```

```
Store the rate limit data in Redis -- we already have a Redis
connection set up.
```

Here is what Claude Code does, step by step.

Step 1: Understand the codebase. Claude Code searches for your route definitions, your Redis configuration, and your existing middleware. It reads several files to understand how your API is structured.

Step 2: Plan the approach. Claude Code describes what it intends to do: create a new rate limiting middleware, wire it into the public routes, and add configuration values. It might ask a clarifying question — "I see you have routes under /api/public and /api/v1/public -- should rate limiting apply to both?"

Step 3: Implement. Claude Code creates the middleware file, writing the sliding window logic using your Redis connection. It follows the patterns it observed in your existing middleware. It asks permission before writing each file.

Step 4: Wire it up. Claude Code modifies your route configuration to apply the new middleware to the appropriate endpoints. It shows you the diff and asks for approval.

Step 5: Add tests. Claude Code creates test files for the rate limiting middleware, using the same testing framework and patterns as your existing tests. It mocks the Redis connection the same way your other tests do.

Step 6: Verify. Claude Code runs your test suite to confirm the new tests pass and the existing tests still pass. If something fails, it reads the error, fixes the issue, and runs the tests again.

Step 7: Report. Claude Code summarizes what it did: which files it created, which files it modified, and what the rate limiting behavior looks like.

The entire process might take two to five minutes, depending on the complexity of your codebase and how many questions Claude Code needs to ask. Compare that to implementing rate limiting manually — reading documentation, writing the middleware, figuring out the Redis data structures, writing tests, debugging failures. Claude Code compresses that process dramatically.

But — and this matters — you should still review the changes. Open the files Claude Code created and modified. Read the rate limiting logic. Make sure the Redis key structure makes sense. Check that the tests cover the edge cases you care about. Claude Code is fast and usually correct, but it is your codebase and your responsibility.

Ending a Session

When you are done, you have several options:

- Type `/exit` to end the session cleanly.
- Press **Ctrl+C** to interrupt and exit.
- Press **Ctrl+D** to send an EOF signal, which also exits.

Before you end a session, run `/cost` one more time to see the total. Over time, this builds a sense of what different types of tasks cost, which informs how you use the tool.

Your conversation history is saved between sessions. When you run `claude` in the same project directory, you can resume a previous conversation or start a new one. Starting fresh is usually better for new tasks — stale context from a previous session can confuse more than it helps.

Lessons from Your First Session

Here are the key takeaways to carry into the rest of this book:

Claude Code is an agent, not a chatbot. It does not just talk about code — it reads, writes, and executes. Every interaction can involve multiple tools and multiple steps.

Orientation matters. The more context Claude Code has about your project, the better its output. Invest in CLAUDE.md files, and spend a moment orienting Claude Code at the start of each session.

Review everything. Claude Code is good, but not perfect. Treat its output the way you would treat a pull request from a colleague — review it, question it, and approve it only when you are satisfied.

The permission model is your friend. It feels like friction at first, but it keeps you in control. Relax permissions gradually as you build trust.

Use /compact and /cost. These two commands keep your sessions efficient and your costs predictable.

The next chapter moves into the first core workflow: code generation. You have seen Claude Code explore a codebase and implement a feature. Now we dig into the techniques and strategies that make code generation with Claude Code consistently excellent.

Part II: Core Development Workflows

"Make it work, make it right, make it fast."

— Kent Beck

Chapter 4: Code Generation

You have Claude Code installed, you have run your first session, and you have a feel for the conversation loop. Now it is time to do real work. Code generation is where most developers first experience the leverage Claude Code provides — not toy snippets, but production-quality code that fits your project, follows your conventions, and runs.

This chapter covers how to generate new files and components, how to get Claude Code to produce multi-file features in a single pass, how to generate tests alongside your production code, and how to lean on Claude Code's awareness of your project structure to keep everything consistent. We close with a real-world walkthrough: building a new API endpoint with tests from scratch.

Generating New Files and Components

The simplest form of code generation is asking Claude Code to create something new. But there is a meaningful difference between asking a chatbot to "write me a React component" and asking Claude Code to do it. Claude Code sits inside your project. It sees your directory layout, your existing components, your imports, your naming conventions, and your configuration files. When you ask it to generate something, it has context a standalone AI never will.

Start with a clear, specific request:

```
Create a new React component called UserProfileCard that displays a user's
avatar, name, email, and join date. It should follow the same patterns as
the existing components in src/components/.
```

Claude Code typically does several things before writing a single line of code. It looks at your existing components to understand your patterns. Are you using functional components with hooks? TypeScript or JavaScript? CSS modules, Tailwind, styled-components? Named exports or default exports? It reads your codebase first, then generates code that fits.

This is the critical difference. You are not getting generic code that you then have to adapt. You are getting code that matches your project from the start.

Being Specific Without Being Prescriptive

There is an art to prompting for code generation. Too vague, and Claude Code makes assumptions you do not want. Too specific, and you are essentially dictating the code yourself, which defeats the purpose.

The sweet spot is describing *what* you want and *why*, while letting Claude Code figure out *how*:

```
I need a rate limiter middleware for our Express app. It should limit each
authenticated user to 100 requests per minute, and unauthenticated requests
to 20 per minute. Use Redis for state since we already have a Redis
connection in src/lib/redis.ts. Return a 429 with a Retry-After header
when the limit is hit.
```

This prompt specifies the requirements (rate limits, Redis, 429 responses) without dictating the implementation. Claude Code can choose the algorithm, structure the code, and integrate it with your existing Redis setup as it sees fit.

Iterating on Generated Code

Code generation is rarely one-shot. You will look at what Claude Code produced and want changes. This is normal. The conversational nature of Claude Code makes iteration natural:

```
This looks good, but can you switch from a fixed-window algorithm to a
sliding-window approach? Also, add a bypass for health check endpoints.
```

Claude Code remembers the full context of what it generated. It modifies the existing code rather than starting over. This iterative loop — generate, review, refine — is the core workflow for code generation.

Generating Code That Fits Your Conventions

Every codebase has conventions, and most of them are unwritten. File naming, import ordering, error handling patterns, logging approaches, variable naming — these conventions live in the code itself rather than in a document.

Claude Code picks up on many of these conventions automatically by reading your existing files. But you can make this much more reliable by documenting your conventions in a `CLAUDE.md` file at the root of your project. With a good CLAUDE.md in place, every piece of code Claude Code generates follows your rules without you repeating them in every prompt. See Chapter 8 for a complete guide to writing effective CLAUDE.md files and templates for different project types.

Multi-File Generation

Real features rarely live in a single file. A new API endpoint might require a route definition, a controller, a service layer, a repository, validation schemas, types, and tests. Claude Code handles this naturally.

When you describe a feature rather than a file, Claude Code thinks in terms of the full feature:

```
Add a new feature for user invitations. A user should be able to invite
others by email. The invitation creates a record in the database, sends an
email via our existing email service, and expires after 48 hours. Follow
the same patterns as the existing user registration flow.
```

Claude Code typically creates multiple files in sequence: a database migration, a model or type definition, a repository, a service, a controller, route registration, and possibly email templates. It creates them in dependency order — types and models first, then the layers that depend on them.

Guiding Multi-File Generation

For complex features, you can guide the structure explicitly:

```
Let's build this in layers. Start with the database migration and types,
then the repository, then the service layer, then the controller and
routes. I want to review each layer before moving to the next.
```

This gives you checkpoints. You review the data layer, confirm it is correct, and Claude Code builds the next layer on a solid foundation. Slower, but better results for complex features, because mistakes in lower layers do not propagate upward.

Alternatively, if you trust the patterns are well-established in your codebase, let Claude Code generate everything at once:

```
Generate all the files for the invitation feature in one pass. Follow the
exact same structure as the user registration feature.
```

Claude Code creates all the files, and you review the full set together.

Test Generation

Test generation is one of Claude Code's strongest capabilities, and it is where many developers see the most immediate time savings.

Generating Tests for Existing Code

Point Claude Code at existing code and ask for tests:

```
Write unit tests for src/services/billing.ts. Cover the happy paths and
the error cases. Use the same testing patterns as the other test files in
this project.
```

Claude Code reads the source file, understands the functions and their signatures, examines your existing test files for patterns (test runner, assertion library, mocking approach, file naming), and generates tests that fit. If your project uses Jest with `describe`/`it` blocks and mock factories, that is what you get. If you use Vitest with test fixtures, you get that instead.

Generating Tests Alongside New Code

Even better, ask for tests as part of the initial generation:

```
Create the invitation service and its unit tests together. I want at least
80% branch coverage.
```

When Claude Code generates code and tests together, the tests are well-matched to the implementation. The tests exercise the code paths that exist rather than testing imagined interfaces.

Test Quality Considerations

A word of caution: Claude Code generates tests that pass. This is both a feature and a risk. Tests written to match an implementation will pass even if the implementation is wrong. They test that the code does what it does, not necessarily what it *should* do.

To counter this, be explicit about expected behavior:

```
Write tests for the invitation service. An invitation to an already-
registered email should fail with a DuplicateUserError. An invitation
from a user who has exceeded their invitation quota (5 per day) should fail
with a QuotaExceededError. Expired invitations should not be redeemable.
```

By specifying expected behaviors, you are writing the spec. Claude Code writes the test code that verifies the spec. This is a much stronger testing approach than "generate tests for this file."

Running Tests to Verify

Claude Code can run your test suite after generating tests. Ask it to:

```
Run the tests you just created and fix any failures.
```

This closes the loop. Claude Code generates tests, runs them, sees failures (maybe an import path is wrong or a mock is misconfigured), and fixes them. You end up with tests that pass against your real codebase.

Project Structure Awareness

Claude Code builds a mental model of your project every time it starts a session. It understands your directory structure, your build configuration, your dependencies, and the relationships between your files.

This awareness shows up in subtle but important ways during code generation:

- **Import paths are correct.** Claude Code knows where files live relative to each other and generates the right import paths, including handling path aliases from your tsconfig or webpack config.
- **Naming follows your patterns.** If your controllers are named `user.controller.ts` and your services are `user.service.ts`, new files follow the same convention.
- **Dependencies are respected.** Claude Code will not import a library you do not have installed, and it uses the version-appropriate API for the libraries you do have.
- **Configuration is honored.** If your ESLint config enforces single quotes and no semicolons, generated code follows those rules.

You can lean into this awareness explicitly:

```
Look at how the users module is structured in src/modules/users/ and
create a new products module that follows exactly the same structure.
```

Claude Code examines every file in the users module, understands the pattern, and replicates it for products — adjusting names, types, and logic to fit the new domain while keeping the structural pattern identical.

When Generation Works Well vs. When It Struggles

Code generation works best when:

- **The patterns are established.** If your codebase has clear, consistent patterns, Claude Code replicates them reliably. The tenth API endpoint is easier to generate than the first.
- **The requirements are clear.** Specific, unambiguous requirements produce specific, correct code.
- **The scope is bounded.** A single component, a single service, a single endpoint — well-scoped generation tasks.
- **The domain is common.** CRUD operations, REST APIs, React components, database queries — Claude Code has deep knowledge of common patterns.

Generation struggles when:

- **The requirements are vague.** "Make the app better" produces nothing useful. "Add a caching layer to the user lookup endpoint with a 5-minute TTL" produces excellent code.
- **The domain is highly specialized.** Niche protocols, uncommon frameworks, or domain-specific algorithms may require more guidance and iteration.
- **The patterns are inconsistent.** If your codebase has three different ways of handling errors, Claude Code has to guess which one you want.
- **The scope is massive.** Asking Claude Code to "build the entire backend" in one prompt does not work. Break large features into manageable pieces.

Real-World Workflow: Building a New API Endpoint with Tests

Let's walk through a realistic workflow. You are working on a Node.js backend with Express, TypeScript, Prisma for the database, and Zod for validation. You need a new endpoint for managing user notification preferences.

Step 1: Describe the Feature

```
I need a new API endpoint for managing user notification preferences.

Requirements:
- GET /api/users/:id/notification-preferences -- returns current preferences
- PUT /api/users/:id/notification-preferences -- updates preferences
- Preferences include: email_notifications (boolean), push_notifications
  (boolean), notification_frequency ("immediate" | "daily" | "weekly"),
  and quiet_hours_start/quiet_hours_end (nullable time strings like "22:00")
- Users can only access their own preferences (enforce via auth middleware)
- If preferences don't exist yet, GET should return defaults
  (all true, frequency "immediate", no quiet hours)

Follow the same patterns as the existing endpoints in src/routes/.
```

Claude Code reads your existing routes, services, and Prisma schema. It understands your project structure and begins generating.

Step 2: Review the Database Layer

Claude Code creates a Prisma migration for the new `NotificationPreference` model and a Zod schema for validation. You review:

```
The migration looks good. One change -- make the notification_frequency
column default to 'immediate' at the database level, not just in the
application code.
```

Claude Code updates the migration.

Step 3: Review the Service and Routes

Claude Code generates the service layer with the business logic (default preferences, ownership checks) and the route handlers. You notice something:

```
The ownership check should use our existing requireOwnership middleware
from src/middleware/auth.ts rather than checking manually in the
controller. Can you refactor to use that?
```

Claude Code knows about that middleware because it already read it. It refactors to use the existing middleware.

Step 4: Generate Tests

```
Now write tests for both the service and the route handlers. For the
routes, use supertest like the other route tests. Mock the database layer.
Make sure to test the default preferences behavior and the ownership
enforcement.
```

Claude Code generates comprehensive tests covering:

- Fetching existing preferences
- Fetching default preferences for a user with none set
- Updating preferences with valid data
- Rejecting invalid frequency values
- Rejecting requests from non-owners
- Handling the quiet hours validation (end must be after start, both or neither must be set)

Step 5: Run and Verify

```
Run the tests and fix any issues. Then run the full test suite to make
sure nothing is broken.
```

Claude Code runs the new tests, fixes a minor import issue, confirms they pass, then runs the full suite to verify no regressions.

Step 6: Final Review

```
Show me a summary of all the files you created or modified.
```

Claude Code lists every file: the Prisma migration, the Zod schema, the service, the route handlers, the route registration, and the test files. You do a final review, and the feature is ready to commit.

The entire workflow took maybe fifteen minutes of your time, most of it spent reviewing and making judgment calls. The mechanical work — writing the migration, the boilerplate, the validation schemas, the test setup — was handled by Claude Code.

Key Takeaways

Code generation with Claude Code is not about getting code for free. It shifts your effort from writing code to directing and reviewing code. You become the architect and reviewer while Claude Code handles the implementation. The quality of the output depends on the quality of your input: clear requirements, established patterns, and iterative refinement produce code you would be proud to ship.

The next chapter looks at the other side of the coin — using Claude Code to review and refactor code that already exists.

Chapter 5: Code Review and Refactoring

Generating new code is exciting. Improving existing code is where the real value often hides. Codebases accumulate complexity over time — rushed features, outdated patterns, inconsistent conventions, and layers of well-intentioned-but-now-confusing abstractions. Claude Code is remarkably effective at reviewing, refactoring, and modernizing existing code, because it can hold large amounts of context and apply changes consistently across many files.

This chapter covers using Claude Code to review your own code and pull requests, refactoring at various scales, performing large-scale changes across an entire codebase, and assisting with migrations. We close with a real-world walkthrough of migrating a codebase from JavaScript to TypeScript.

Using Claude Code to Review Your Own Code

Every developer knows the experience of writing code in a focused state and then, the next morning, spotting obvious problems. Claude Code gives you that fresh-eyes perspective immediately.

After writing a feature, ask Claude Code to review it:

```
Review the changes I've made in this branch compared to main. Look for
bugs, edge cases I might have missed, performance issues, and anything
that doesn't follow our project conventions.
```

Claude Code runs `git diff main`, reads every changed file, and provides a structured review. A typical response might identify:

- A missing null check on a value that could be undefined
- An N+1 query in a loop that should be batched
- An error that is caught and silently swallowed instead of logged
- A new function that does not match the naming convention used elsewhere in the module

This does not replace a human code review from a teammate. It is a pre-review — a way to catch the easy stuff before your colleague ever sees the PR. Many developers find that running Claude Code review before opening a PR eliminates the most common review comments and lets human reviewers focus on the design and architectural questions that matter more.

Targeted Reviews

You can focus Claude Code's review on specific concerns:

```
Review src/services/payment.ts for security issues. I'm handling credit
card data and want to make sure I'm not logging sensitive information or
exposing it in error responses.
```

Or ask about specific quality dimensions:

```
Review the error handling in the new checkout flow. Are all the error
cases handled? Are the error messages useful for debugging without leaking
internal details to the client?
```

Targeted reviews are often more useful than broad "review everything" requests, because Claude Code goes deeper on the dimension you care about.

Reviewing Pull Requests and Diffs

Claude Code can review pull requests directly. If you are using GitHub, you can point it at a PR:

```
Review the pull request at https://github.com/our-org/our-repo/pull/247.
Summarize what the PR does and flag any concerns.
```

Claude Code fetches the PR details, reads the diff, and provides a review. This is especially useful for:

- **Large PRs** that are hard to review file by file. Claude Code can hold the entire diff in context and identify cross-file issues that are hard to spot when reviewing individual files.
- **Unfamiliar areas of the codebase.** If a PR touches code you do not work with often, Claude Code can explain what the code does and whether the changes make sense.

- **Consistency checks.** Does the PR follow the patterns established elsewhere in the codebase? Claude Code compares the new code against existing code and flags deviations.

Reviewing Your Own Diffs Before Committing

A quick pre-commit review is one of the highest-value habits you can build:

```
Look at my staged changes. Anything I should fix before committing?
```

Claude Code reads the staged diff and gives you a quick sanity check. This catches typos, debugging artifacts left in the code, missing imports, and other small issues that are embarrassing to commit.

Refactoring: Renaming, Extracting, Restructuring

Refactoring is the bread and butter of codebase maintenance. Claude Code handles the full spectrum, from simple renames to complex restructuring.

Renaming

Renaming seems simple, but doing it correctly across a codebase is surprisingly tedious. A variable rename might touch dozens of files. A function rename requires updating every call site, every import, every test, and every doc reference.

```
Rename the function getUserData to fetchUserProfile everywhere in the
codebase. Update all call sites, imports, tests, and comments.
```

Claude Code searches the entire codebase for references, updates them all, and shows you what it changed. This goes beyond simple find-and-replace — Claude Code understands context. It will not rename a `getUserData` in an unrelated third-party type definition, and it updates destructured imports correctly.

Extracting

Extraction refactors — pulling a function, a class, or a module out of a larger piece of code — are where Claude Code saves significant time:

```
The processOrder function in src/services/orders.ts is 200 lines long.
Extract the discount calculation logic into its own function, and extract
the inventory check into its own function. Keep them in the same file for
now.
```

Claude Code reads the function, identifies the logical boundaries, extracts the pieces, wires up the parameters and return values, and updates the original function to call the extracted pieces. The resulting code does exactly the same thing but is far easier to read and test.

Restructuring

More ambitious restructuring is also possible:

```
Our utils.ts file has grown to 800 lines and contains unrelated utilities.
Split it into logical modules: string-utils.ts, date-utils.ts,
validation-utils.ts, and http-utils.ts. Update all imports across the
codebase.
```

Claude Code reads the file, categorizes each function, creates the new modules, moves the functions, updates every import in the codebase, and verifies nothing is broken. What would take you an hour of careful, tedious work takes Claude Code a few minutes.

Large-Scale Refactoring Across Many Files

This is where Claude Code distinguishes itself from simpler tools. Some refactoring tasks require coordinated changes across dozens or hundreds of files. Manual approaches are error-prone and exhausting. Simple find-and-replace is too blunt.

Pattern-Based Refactoring

You can describe a pattern transformation and let Claude Code apply it everywhere:

```
We're switching from our custom Result type to neverthrow's Result type
across the codebase. In every file that uses Result<T, E>, update the
imports to use neverthrow, and update any pattern matching from our custom
.match() syntax to neverthrow's .match() syntax. The main difference is
that neverthrow's match takes an object with ok and err methods instead
of two positional callbacks.
```

Claude Code finds every file using the old pattern, understands the semantic difference between the two APIs, and makes the correct transformation in each case. It handles variations — some files use `.match()`, some use `.isOk()`

and `.isErr()`, some destructure the result. Each instance gets the appropriate transformation.

API Changes

When a library updates its API, Claude Code can update your codebase to match:

```
We're upgrading from React Router v5 to v6. Update all route definitions,
replace Switch with Routes, replace component match props with useParams,
replace Redirect with Navigate, and replace useHistory with useNavigate. Here's
the migration guide:
[paste relevant sections]
```

You can paste in the migration guide or describe the changes. Claude Code applies them across every file that needs updating. For library migrations with well-known patterns, Claude Code often already knows the transformation rules without needing a guide.

Incremental Large-Scale Changes

For very large refactors, consider an incremental approach:

```
Let's convert our error handling to the new pattern one module at a time.
Start with the users module. Show me the changes before applying them.
```

Review the changes for one module, confirm they are correct, then move to the next. This gives you confidence that the transformation is correct before applying it at scale.

Migration Assistance

Migrations — framework upgrades, language version bumps, dependency replacements — are among the most dreaded tasks in software development. They are large, they touch everything, and they are easy to get subtly wrong. Claude Code makes migrations dramatically less painful.

Framework Upgrades

```
We need to upgrade from Next.js 13 to Next.js 14. The main changes are
the new App Router patterns and the server/client component distinction.
Walk me through what needs to change in our codebase.
```

Claude Code scans your codebase, identifies every file affected by the upgrade, and either makes the changes or provides a prioritized plan for doing so.

Dependency Replacements

```
Replace moment.js with date-fns throughout the codebase. Moment is
deprecated and we want to move to something tree-shakable. Map each
moment function we use to its date-fns equivalent.
```

Claude Code finds every moment usage, determines the equivalent date-fns function, updates the imports and calls, and adjusts for API differences (like date-fns using plain Date objects instead of moment's wrapper objects).

Configuration Migrations

Sometimes migrations are mostly about configuration — moving from one build tool to another, one linter to another, one testing framework to another:

```
Migrate our Jest configuration to Vitest. Update jest.config.ts to
vitest.config.ts, update the test scripts in package.json, and handle
any API differences in our test files (jest.fn() to vi.fn(), etc.).
```

Claude Code handles both the configuration changes and the source code changes that follow from them.

Real-World Workflow: Migrating a Codebase from JavaScript to TypeScript

Let's walk through one of the most common and impactful migrations: converting a JavaScript project to TypeScript. This is a substantial effort that touches every file and requires careful attention to avoid breaking a working application.

Step 1: Assess the Current State

```
Analyze this JavaScript project and tell me: how many JS files are there,
what frameworks and libraries are used, is there any existing TypeScript
configuration, and are there any files that will be particularly difficult
to convert?
```

Claude Code scans the entire project and reports back. Maybe it finds 85 JavaScript files, an Express backend with Sequelize, a React frontend, no existing

TypeScript config, and a few files with complex dynamic patterns that will need careful typing.

Step 2: Set Up TypeScript

```
Set up TypeScript for this project. Add tsconfig.json with strict mode
enabled but allowJs true so we can migrate incrementally. Install the
necessary type packages (@types/express, @types/node, etc.) for our
dependencies. Set up the build scripts.
```

Claude Code creates the configuration, installs dependencies, and configures the build pipeline. You now have a project that can run a mix of JS and TS files.

Step 3: Migrate Incrementally

Rather than converting everything at once, you work module by module:

```
Convert the src/models/ directory from JavaScript to TypeScript. Add proper
types for all model properties. Infer types from how the models are used
elsewhere in the codebase.
```

Claude Code converts each file, inferring types from existing usage patterns. A `user.js` that creates objects with `name`, `email`, and `createdAt` gets proper types for each field. Claude Code examines how these models are used in services and controllers to make sure the types are accurate and complete.

```
Good. Now convert src/services/ to TypeScript. Use the types from the
models we just converted.
```

Each layer builds on the previous one. By the time you reach the top-level route handlers, the types from the models and services are already in place.

Step 4: Handle the Hard Parts

Some files are harder to type than others:

```
The src/lib/query-builder.js file uses a lot of dynamic patterns. Convert
it to TypeScript and add the best types you can. Where the types are
genuinely dynamic, use generics or branded types rather than `any`.
```

Claude Code works through the difficult cases, using generics, conditional types, or discriminated unions as appropriate. Where typing is impractical, it uses explicit `unknown` with type guards rather than sprinkling `any` throughout the code.

Step 5: Verify

```
Run the TypeScript compiler with --noEmit and show me any remaining type
errors. Let's fix them together.
```

Claude Code runs `tsc --noEmit`, reads the errors, and fixes them one by one. Some are simple mistakes. Others reveal real bugs in the original JavaScript code — places where the wrong type was being passed or where a null value was not being handled. This is one of the hidden benefits of TypeScript migration: it finds bugs.

Step 6: Clean Up

```
Remove the allowJs flag from tsconfig.json, delete any remaining .js files
that have been converted, and run the full test suite to make sure
everything still works.
```

Claude Code tightens the TypeScript configuration, cleans up artifacts, and runs the tests. The migration is complete.

What Made This Manageable

The key to this workflow is the incremental approach. You did not ask Claude Code to convert 85 files in one shot. You worked layer by layer, reviewing as you went, building confidence with each step. Claude Code maintained context across the entire conversation, so types defined in step 3 were used correctly in step 4.

Key Takeaways

Code review and refactoring with Claude Code lets you maintain and improve your codebase at a pace that would be impractical manually. Pre-commit reviews catch bugs before they reach your teammates. Refactoring becomes something you do rather than something you add to the backlog and never touch. Migrations go from multi-week projects to multi-day efforts.

The pattern is consistent: describe the change you want, let Claude Code do the mechanical work, review and refine, and verify with your existing test suite. You stay in the driver's seat. Claude Code handles the repetitive execution.

The next chapter puts Claude Code to work on one of development's most frustrating tasks: debugging.

Chapter 6: Debugging

Debugging is where many developers waste the most time — not because they lack skill, but because debugging requires holding enormous amounts of context in your head at once. You need to understand the expected behavior, the actual behavior, the code paths involved, the state at each step, and the interactions between components. Claude Code excels at exactly this kind of multi-dimensional analysis. It reads your code, your logs, your error messages, and your test output, then synthesizes all of it into a diagnosis.

This chapter covers how to describe bugs effectively to Claude Code, how to leverage error messages and stack traces, debugging strategies that work well with an AI assistant, debugging across multiple files and layers, and performance debugging. We close with a real-world walkthrough of tracking down a race condition.

Describing Bugs to Claude Code Effectively

The quality of Claude Code's debugging assistance depends on the quality of your bug description. Vague descriptions produce vague suggestions. Precise descriptions produce precise diagnoses.

The Good Bug Report Format

A good bug description for Claude Code has three parts:

1. **What should happen.** The expected behavior, as specifically as possible.
2. **What actually happens.** The actual behavior, including exact error messages.
3. **When it happens.** The conditions under which the bug appears.

Here is a weak description:

```
The login is broken.
```

And here is a strong one:

```
When a user logs in with valid credentials, they should be redirected to
/dashboard. Instead, they're redirected to /login again with no error
message. This only happens when the user has two-factor authentication
enabled. Users without 2FA can log in normally. The server logs show the
authentication succeeds (200 response from /api/auth/login) but the
redirect after 2FA verification goes to the wrong place.
```

The second description gives Claude Code enough context to start narrowing down the problem immediately. It knows to look at the 2FA verification flow, not the initial login. It knows authentication itself works. It knows the issue is in the redirect logic after 2FA.

Let Claude Code Explore

You do not always have a detailed bug description. Sometimes all you know is "it is broken." Fine. Start with what you know and let Claude Code investigate:

```
Users are reporting that the dashboard sometimes shows stale data. I can't
reproduce it reliably, but it seems to happen more often right after a
deployment. The relevant code is in src/pages/dashboard/ and
src/services/dashboard-data.ts.
```

Claude Code reads those files, traces the data flow, looks for caching logic, checks for race conditions between data loading and rendering, and forms hypotheses. It might ask follow-up questions, or it might identify a likely cause immediately.

Having Claude Code Read Error Messages and Stack Traces

Error messages and stack traces are gold for debugging, and Claude Code is exceptionally good at interpreting them. Do not summarize your error messages — paste them in full.

```
I'm getting this error in production:

TypeError: Cannot read properties of undefined (reading 'preferences')
    at UserSettingsService.getNotificationPrefs (dist/services/user-settings.js:
47:38)
    at async DashboardController.loadDashboard (dist/controllers/dashboard.js:23:
22)
    at async Layer.handle [as handle_request] (/app/node_modules/express/lib/rou
ter/layer.js:95:5)

Here's the relevant source code path: src/services/user-settings.ts
```

Claude Code reads the stack trace, maps it to your source code (handling the dist/src difference), reads the relevant file, and identifies the issue. It might find that `getNotificationPrefs` assumes the user record always has a `settings` property, but new users created through a recently added OAuth flow do not have one yet.

Multiple Error Messages

Sometimes the real bug is not where the error occurs:

```
I'm seeing these errors intermittently in our logs. They seem related but
I'm not sure which is the root cause:

[Error 1]: Connection timeout to Redis at 10.0.0.5:6379
[Error 2]: Session lookup failed: ECONNREFUSED
[Error 3]: TypeError: Cannot read properties of null (reading 'userId')

These happen roughly at the same time, maybe a few seconds apart.
```

Claude Code can correlate these errors and reason about causality. The Redis connection times out, which causes the session lookup to fail, which returns null instead of a session object, which causes the null reference error. The root cause is the Redis connection, not the TypeError.

Debugging Strategies

Claude Code is not just an error message interpreter. It can debug alongside you using several strategies.

Log-Based Debugging

```
I can't figure out why the order total is sometimes calculated wrong. Can
you add strategic logging to the order calculation pipeline in
src/services/order-calculator.ts so we can trace the values through each
step? Use our standard logger.
```

Claude Code reads the calculation flow, identifies key decision points and transformations, and adds logging at each stage. You run the scenario, collect the logs, and paste them back:

```
Here are the logs from a case where the total was wrong:

[OrderCalc] Subtotal: 99.99
```

```
[OrderCalc] Discount (SUMMER20): -20.00
[OrderCalc] After discount: 79.99
[OrderCalc] Tax rate: 0.0875
[OrderCalc] Tax amount: 7.00
[OrderCalc] Shipping: 5.99
[OrderCalc] Total: 92.99

The customer expected $92.98. We're off by a penny.
```

Claude Code reads the logs and spots it: the tax is calculated as `0.0875 * 79.99 = 6.999125`, which gets rounded to `7.00` at the final total but the intermediate value `6.999125` is carried forward and combined with shipping before rounding. The fix is to round each line item to the nearest cent before summing, not round only the final total.

Hypothesis-Driven Debugging

Claude Code can form and test hypotheses:

```
The webhook handler sometimes processes the same event twice. I suspect
it's a race condition but I'm not sure. Can you look at the webhook
handler in src/webhooks/stripe.ts and the idempotency check in
src/lib/idempotency.ts and tell me if there's a window where duplicates
could get through?
```

Claude Code reads both files, traces the execution flow, and analyzes the timing. It might find that the idempotency check reads from the database, but the idempotency key is not written until after the webhook is processed. If two identical webhook deliveries arrive within milliseconds, both can pass the idempotency check before either writes its key.

Bisection

For bugs that appeared at some unknown point, Claude Code can help with git bisect:

```
The search feature stopped returning results for quoted phrases at some
point in the last two weeks. Can you help me set up a git bisect to find
when it broke? The test case is: searching for "exact phrase" should
return results that contain that exact phrase.
```

Claude Code can write a bisect script, help you interpret bisect results, and once the offending commit is found, analyze the diff to explain what change caused the regression.

Debugging Across Multiple Files and Layers

Many bugs do not live in a single function. They live in the interaction between components — a value correctly computed in one layer but incorrectly passed to the next, a type coercion that happens silently at an API boundary, a race condition between two independent systems.

Claude Code's ability to read and reason about multiple files at once makes it strong at these cross-cutting bugs:

```
We have a bug where user preferences saved through the mobile API don't
show up in the web app. Both use the same database. The mobile API is in
src/api/mobile/, the web API is in src/api/web/, and the preferences
service is in src/services/preferences.ts. Can you trace the data flow
for saving and loading preferences through both paths and find where they
diverge?
```

Claude Code reads all the relevant files, traces the save path through the mobile API and the load path through the web API, and identifies the divergence. Maybe the mobile API saves preferences with a different key format. Maybe the web API filters by a tenant ID that the mobile API does not set. Whatever the issue, Claude Code sees both paths at once and compares them.

Layer-by-Layer Investigation

For complex bugs, work from the outside in:

```
Start at the HTTP layer. Show me what request the client is sending and
what response the server returns for this failing case.
```

Then:

```
OK, the response is a 200 with an empty array. Trace backwards from the
controller through the service to the database query. Where does the data
disappear?
```

Then:

```
The database query returns 5 rows. So the problem is between the database
and the controller. What happens to the data in the service layer?
```

This methodical approach, narrowing the search space at each step, is how experienced debuggers work. Claude Code is a patient partner for this process.

Performance Debugging

Performance issues are a distinct category of debugging. The code is correct — it just takes too long or uses too many resources.

Identifying Bottlenecks

```
The /api/dashboard endpoint takes 3-4 seconds to respond. It was under
200ms a month ago. Can you look at the endpoint and identify what might
be slow? Check for N+1 queries, unnecessary data loading, missing
database indexes, and any synchronous operations that should be async.
```

Claude Code reads the endpoint, traces the data loading, and identifies performance issues. Common findings include:

- An ORM query that loads related records one at a time instead of batch loading
- A database query that performs a full table scan because an index is missing
- Data transformations that iterate over the same array multiple times
- API calls to external services that could be parallelized but are awaited sequentially

Analyzing Profiler Output

If you have profiling data, Claude Code can interpret it:

```
I ran the Node.js profiler on our server. Here are the top entries from
the CPU profile:

[paste profiler output]

What's taking so much time and what can we do about it?
```

Claude Code reads profiler output, identifies hot paths, and suggests optimizations — algorithmic improvements, caching, or restructuring to avoid expensive operations.

Memory Debugging

```
Our Node.js server's memory usage grows over time and eventually OOMs.
I took a heap snapshot at startup and another after 4 hours. The biggest
growth is in (string) and (array). Can you look at our event handling
code in src/events/ for potential memory leaks? Specifically check for
event listeners that aren't cleaned up.
```

Claude Code examines the code for common memory leak patterns: event listeners added in loops without corresponding removal, closures that capture large objects, caches without eviction, and growing arrays that are never trimmed.

Real-World Workflow: Tracking Down a Race Condition

Race conditions are among the hardest bugs to debug because they are non-deterministic, they often only manifest under load, and the code looks correct when you read it sequentially. Let's walk through debugging one with Claude Code.

The Bug

You have a ticket-purchasing system. Users are occasionally buying the same ticket simultaneously — the last ticket for a popular event sold to two different users. Your code is supposed to prevent this.

Step 1: Describe the Problem

```
We have a race condition in our ticket purchasing system. The same ticket
is sometimes sold to two different users. This happens maybe once every
few hundred purchases, always for events that are nearly sold out.

The relevant code is:
- src/services/ticket-purchase.ts -- the main purchase flow
- src/repos/ticket-repo.ts -- database access for tickets
- src/repos/event-repo.ts -- database access for events

The purchase flow checks available inventory, reserves a ticket, charges
the user, and confirms the purchase. We're using PostgreSQL.
```

Step 2: Claude Code Analyzes the Code

Claude Code reads all three files and identifies the core issue. The purchase flow looks something like:

1. Query: How many tickets remain for this event?
2. If tickets remain, proceed.
3. Insert a ticket_purchase record.
4. Charge the user.
5. Mark the ticket as sold.

Claude Code spots the problem: there is a gap between step 1 (checking availability) and step 3 (claiming a ticket). Two concurrent requests can both check availability, both see one ticket remaining, and both proceed to purchase it.

Step 3: Understand the Scope

```
Is this the only place where ticket availability is checked? Could the
same race exist in any other code path?
```

Claude Code searches the codebase for other places that query ticket availability and finds a similar pattern in the waitlist-to-purchase flow. The same race condition exists there.

Step 4: Discuss Solutions

```
What are our options for fixing this? I want something robust that handles
high concurrency. We're on PostgreSQL.
```

Claude Code proposes several approaches with trade-offs:

- **Database-level locking:** Use `SELECT ... FOR UPDATE` to lock the inventory row during the transaction. Simple and reliable, but serializes all purchases for an event.
- **Optimistic locking:** Add a version column to the events table. Check the version when decrementing inventory. If it has changed, retry. Good for moderate concurrency.
- **Atomic decrement:** Use a single `UPDATE events SET tickets_remaining = tickets_remaining - 1 WHERE id = $1 AND`

`tickets_remaining > 0` query. If the update affects zero rows, the tickets are sold out. No separate check needed.

Claude Code recommends the atomic decrement approach for this case: simple, fast, and eliminates the race condition entirely by making the check and the decrement a single atomic operation.

Step 5: Implement the Fix

```
Implement the atomic decrement approach. Update both the regular purchase
flow and the waitlist-to-purchase flow. Make sure the charge only happens
after we've successfully decremented the inventory.
```

Claude Code rewrites the purchase flow:

1. Begin transaction.
2. Atomic decrement: `UPDATE events SET tickets_remaining = tickets_remaining - 1 WHERE id = $1 AND tickets_remaining > 0 RETURNING tickets_remaining`.
3. If no row returned, abort — sold out.
4. Insert the ticket_purchase record.
5. Charge the user.
6. Commit transaction (or roll back if the charge fails, which re-increments the count).

It applies the same pattern to the waitlist flow.

Step 6: Write a Regression Test

```
Write a test that verifies the race condition is fixed. Simulate multiple
concurrent purchase attempts for the last ticket and verify that exactly
one succeeds.
```

Claude Code writes a test that uses `Promise.all` to fire multiple purchase requests simultaneously, then asserts that exactly one succeeds with a ticket and the rest fail with a sold-out error. It also verifies the event's remaining ticket count is zero, not negative.

Step 7: Verify

```
Run the new test and the full test suite.
```

The test passes. The race condition is gone. The full suite confirms nothing else is broken.

What Made This Effective

The debugging process worked because it followed a systematic approach: describe the problem, let Claude Code analyze the code and identify the root cause, discuss solutions with trade-offs, implement the fix, and verify it with a targeted regression test. At no point did you need to hold the entire transaction flow in your head — Claude Code held that context for you.

Key Takeaways

Debugging with Claude Code is most effective when you give it complete information — full error messages, exact reproduction steps, relevant file paths — and then let it explore and reason. Work from symptoms to root causes. Use Claude Code's ability to read multiple files at once to trace bugs across layers. For performance issues, point Claude Code at specific endpoints or operations and let it identify bottlenecks.

The biggest shift is psychological. Instead of staring at code and trying to hold everything in your head, you describe the problem to Claude Code and have a conversation about it. The debugging process becomes collaborative rather than solitary.

The next chapter turns Claude Code's analytical abilities toward a different kind of writing: documentation.

Chapter 7: Documentation and Technical Writing

Documentation is the perpetual second-class citizen of software development. Everyone agrees it matters. Almost no one enjoys writing it. The result is predictable: codebases with brilliant implementations and virtually no documentation, or documentation that was accurate two years ago and has been wrong ever since.

Claude Code changes the economics of documentation. What used to take hours of tedious writing takes minutes of prompting and review. More importantly, Claude Code generates documentation directly from your source code, which means it is accurate by construction rather than by good intentions.

This chapter covers generating READMEs, API documentation, Architecture Decision Records, inline comments and docstrings, migration guides, changelogs, and technical blog posts. We close with a real-world walkthrough of documenting an undocumented codebase from scratch.

Generating README Files

A good README is the front door to your project. It tells a new developer what the project is, how to set it up, how to use it, and where to find more information. Claude Code can generate a README that actually covers these bases because it reads your entire project to understand what it does.

```
Generate a README.md for this project. Include: a brief description of
what the project does, prerequisites, installation steps, how to run it
locally, how to run tests, the project structure, and how to contribute.
Base everything on the actual project -- don't make anything up.
```

Claude Code reads your package.json (or equivalent), your scripts, your directory structure, your configuration files, and your existing documentation (if any). It produces a README grounded in reality. The installation steps reflect your actual dependencies. The "how to run" section uses your actual scripts. The project structure section matches your actual directories.

Keeping READMEs Updated

A README generated from source code today will be out of date tomorrow. The key to keeping READMEs current is making updates cheap enough that they happen:

```
The README is out of date. Compare it against the current project state
and update any sections that are no longer accurate. Don't rewrite
sections that are still correct.
```

Claude Code diffs the README against the current project and makes targeted updates. New environment variables that need documenting, scripts that have changed, dependencies that have been added — Claude Code catches the drift.

API Documentation from Source Code

API documentation is simultaneously the most important and most neglected form of documentation. Your API is a contract with your consumers, and undocumented APIs are barely usable APIs.

REST API Documentation

```
Generate API documentation for all the endpoints in src/routes/. For each
endpoint, document: the HTTP method and path, required and optional
parameters, the request body schema with types, the response schema with
example responses, authentication requirements, and possible error
responses. Format it as Markdown.
```

Claude Code reads every route file, traces through to the controllers and validation schemas, and produces comprehensive endpoint documentation. If you use Zod or Joi for validation, Claude Code reads the schemas to pull exact field types and constraints. If you have TypeScript types for your responses, those become the response documentation.

OpenAPI/Swagger Generation

If you want machine-readable API docs:

```
Generate an OpenAPI 3.0 spec for our API based on the route definitions
and Zod schemas. Output it as openapi.yaml.
```

Claude Code translates your route definitions and validation schemas into a formal OpenAPI specification. This spec then drives tools like Swagger UI, API testing tools, and client code generators.

Library API Documentation

For libraries and SDKs, documentation means documenting your public interface:

```
Generate API documentation for the public exports of this library. For
each exported function, class, and type, document: the signature, a
description of what it does, parameter descriptions with types, return
value description, example usage, and any exceptions it can throw.
```

Claude Code reads your exports, understands the types, and generates documentation that covers the public surface. It infers what functions do from their implementation and from how they are used in tests.

Architecture Decision Records (ADRs)

ADRs capture the *why* behind your architectural decisions. They are invaluable when someone (including future you) asks "why did we build it this way?" Claude Code is surprisingly good at helping write ADRs because it can examine the code to see what was decided, and you provide the context for why.

```
Help me write an ADR for our decision to use event sourcing for the order
management system instead of a traditional CRUD approach. We made this
decision because:
- Order state transitions are complex and need an audit trail
- We need to support retroactive corrections
- The business wants analytics on order lifecycle patterns

Use the format: Title, Status, Context, Decision, Consequences.
```

Claude Code drafts the ADR with the technical context filled in from the codebase (what the order system looks like, what event sourcing means for this specific implementation) and the business context from your description. The result is a well-structured document future developers will actually find useful.

Retrospective ADRs

For decisions that were made long ago without documentation, Claude Code can help reconstruct the reasoning:

```
We're using Redis for session storage instead of our PostgreSQL database.
There's no documentation on why. Can you look at the session
implementation and help me write a retrospective ADR? I think it was for
performance reasons, but you might see other factors in the code.
```

Claude Code examines the session implementation, spots patterns like TTL-based expiration, cluster-aware key distribution, and pub/sub for session invalidation, and helps you reconstruct and document the reasoning that likely drove the decision.

Inline Code Comments and Docstrings

There are two schools of thought on inline comments. One says code should be self-documenting and comments are a smell. The other says complex logic deserves explanation. Wherever you fall on this spectrum, Claude Code can help.

Adding Comments to Complex Code

```
Add explanatory comments to the functions in src/lib/rate-limiter.ts. The
algorithm is non-obvious and new team members struggle with it. Don't
comment the obvious stuff -- focus on the *why* and the algorithm logic.
```

Claude Code reads the code, understands the algorithm (maybe a sliding-window rate limiter with token bucket overflow), and adds comments that explain the non-obvious parts. The comment on a bit-shift operation explains what it computes. The comment before a conditional explains the edge case it handles.

Generating Docstrings

For languages with docstring conventions (Python, TypeScript with JSDoc, Java with Javadoc), Claude Code can add docstrings systematically:

```
Add JSDoc comments to all exported functions in the src/utils/ directory.
Include @param, @returns, and @throws tags. Write descriptions that
explain what the function does, not just restate the function name.
```

The distinction in that last sentence matters. A docstring that says `@description Gets the user` for a function called `getUser` is useless. Claude Code writes docstrings that add real information: what parameters are expected, what edge cases exist, what the return value represents.

Batch Docstring Generation

For a large codebase that lacks docstrings entirely:

```
Add JSDoc comments to all exported functions across the entire
src/services/ directory. Skip functions that already have JSDoc. Prioritize
accuracy -- if you're not sure what a function does, add a TODO comment
instead of guessing.
```

Claude Code works through each file systematically, adding docstrings based on its analysis of each function's implementation and usage. The instruction to use TODO for uncertain cases prevents hallucinated documentation, which is worse than no documentation.

Migration Guides and Changelogs

When you make breaking changes to an API or library, your users need to know what changed and how to update. Claude Code can generate these documents directly from your code changes.

Migration Guides

```
We're releasing v3 of our SDK, which has breaking changes compared to v2.
Compare the public API of src/v2/ with src/v3/ and generate a migration
guide. For each breaking change, show the v2 code pattern and the
equivalent v3 code pattern.
```

Claude Code diffs the two API surfaces, identifies every breaking change, and generates a migration guide with before/after code examples. Far more reliable than trying to remember every change you made over the course of development.

Changelogs

```
Generate a changelog entry for the upcoming release. Look at all commits
since the last tag (v2.4.0) and group them into: features, bug fixes,
breaking changes, and internal improvements. Use conventional commit
format. Link to the relevant PRs where possible.
```

Claude Code reads the git log, categorizes the commits, and produces a structured changelog. It groups related commits (three commits that together implement one feature get listed as a single feature entry) and flags breaking changes prominently.

Keeping a Changelog Current

For projects that maintain a running changelog file:

```
Update CHANGELOG.md with the changes from this PR. Add entries under the
Unreleased section. Determine the appropriate category (Added, Changed,
Fixed, Removed) for each change.
```

This becomes a natural part of your PR workflow. Every PR includes its changelog entry, generated by Claude Code and reviewed by you.

Technical Blog Posts

You just built something interesting. Your team lead wants a blog post about it. You would rather write code. Claude Code can draft technical blog posts based on your implementation work.

```
Write a technical blog post about the real-time notification system we
just built. The code is in src/services/notifications/ and
src/websocket/. Target audience is intermediate backend developers.
Cover the architecture, the key design decisions (why WebSockets instead
of SSE, why Redis pub/sub for fan-out), and include code snippets for
the interesting parts. Tone should be conversational and practical.
```

Claude Code reads the implementation, understands the architecture, and produces a blog post that explains it to an external audience. The code snippets are pulled from your actual code (simplified where appropriate). The explanations are grounded in what you built, not in theoretical abstractions.

From Implementation to Explanation

Claude Code is particularly good at translating implementation details into accessible explanations:

```
I need to explain our CQRS implementation to the team. Can you write up a
technical explanation based on the code in src/commands/ and src/queries/?
Assume the audience knows general backend development but hasn't worked
with CQRS before. Use our actual code as examples.
```

The result is an explanation rooted in your specific implementation — far more useful for your team than a generic CQRS article from the internet.

Real-World Workflow: Documenting an Undocumented Codebase

Let's walk through the most common documentation scenario: a codebase that works, that people ship features in, and that has essentially zero documentation. A new developer joins the team and cannot figure out how anything works without reading every file. You have been asked to fix this.

Step 1: Get the Lay of the Land

```
Analyze this codebase and give me a high-level overview. What does it do,
what technologies does it use, how is it structured, and what are the main
modules? Don't go deep yet -- just the bird's eye view.
```

Claude Code scans the project and produces something like: "This is a multi-tenant SaaS application for project management. Built with Node.js, Express, TypeScript, Prisma with PostgreSQL, and React with Vite for the frontend. The backend is organized into modules under `src/modules/` (auth, projects, tasks, billing, notifications). The frontend is in `client/` with a standard React Router setup."

This overview alone is more documentation than the project had before.

Step 2: Generate the Project README

```
Generate a comprehensive README.md. Include: project description,
architecture overview, prerequisites (with versions), setup instructions,
how to run locally, how to run tests, project structure with descriptions
of each top-level directory, environment variables needed (check .env.example),
and deployment notes. Base everything on what you find in the codebase.
```

Claude Code produces a thorough README. You review it, correct a few details (maybe the Docker setup has an extra step not obvious from the code), and commit it.

Step 3: Document the Architecture

```
Create docs/architecture.md. Document: the overall architecture (client,
API, database, Redis, external services), the module structure and how
modules interact, the request lifecycle (from HTTP request to database
and back), the authentication flow, and the multi-tenancy approach. Use
diagrams in Mermaid syntax where they'd help.
```

Claude Code traces through the codebase and produces an architecture document with Mermaid diagrams showing the request flow, the module dependencies, and the authentication sequence. This is the document that helps new developers build a mental model of the system.

Step 4: Document the API

```
Generate docs/api.md with documentation for every API endpoint. Group
endpoints by module. For each endpoint, document the method, path,
authentication requirements, request parameters and body, response
schema, and error cases. Pull the schemas from the Zod validation files.
```

Claude Code reads every route file and validation schema and produces comprehensive API documentation. For a project with fifty endpoints, this saves days of work.

Step 5: Document the Data Model

```
Generate docs/data-model.md from the Prisma schema. For each model,
document: what it represents, its fields with types and descriptions,
its relationships to other models, and any important constraints or
indexes. Also include an entity-relationship diagram in Mermaid.
```

Claude Code reads `schema.prisma` and produces a human-readable description of every table, every relationship, and the overall data model. The Mermaid ER diagram gives a visual overview the raw schema file cannot.

Step 6: Document Key Workflows

```
Create docs/workflows/ with a document for each of these key business
workflows:
- User registration and onboarding
- Creating and managing projects
- Task lifecycle (creation to completion)
- Billing and subscription management
- Notification delivery

For each workflow, trace the code path from the user action through the
API to the database, noting which modules and services are involved.
```

Claude Code traces each workflow through the codebase and produces step-by-step documentation of how each process works. These workflow documents are what developers reach for when they need to modify a feature.

Step 7: Add Inline Documentation

```
Add JSDoc comments to all exported functions in src/modules/billing/.
This is the module new developers struggle with most. Focus on explaining
the business logic and the edge cases.
```

You repeat this for the modules that are most complex or most frequently modified. You do not need docstrings on every function in the codebase on day one — start with the areas that cause the most confusion.

Step 8: Create a Development Guide

```
Create docs/development.md covering: how to set up a local development
environment from scratch, common development tasks (adding a new endpoint,
adding a new database migration, running specific test suites), debugging
tips specific to this project, and the deployment process.
```

This is the document that turns a new developer's first week from five days of confusion into one day of following a guide.

The Result

In a few hours of prompting and reviewing, you have gone from zero documentation to:

- A comprehensive README
- An architecture document with diagrams
- Complete API documentation
- A data model reference
- Workflow documentation for key business processes
- Inline documentation for the most complex modules
- A development guide for new team members

None of these documents are perfect. They will need refinement as people use them and find gaps. But they are 80–90% accurate because they were generated from the actual codebase, which beats the nothing most teams have today.

Maintaining the Documentation

Documentation rots. Code changes; documentation does not. The solution is to make documentation updates a natural part of your development workflow:

```
I just added a new notification channel (SMS). Update the relevant
documentation: the API docs for the new endpoints, the notification
delivery workflow document, the architecture doc if the new SMS provider
changes anything, and the README if there are new environment variables.
```

Because Claude Code reads both the code changes and the existing documentation, it makes targeted updates rather than requiring a full rewrite. Keeping documentation current becomes a two-minute task at the end of a feature rather than a project in itself.

Key Takeaways

Documentation with Claude Code is different from writing documentation by hand. You are not staring at a blank page trying to describe your system. You are pointing Claude Code at your code and asking it to explain the code in various formats for various audiences. The documentation is grounded in reality because it is generated from reality.

The most impactful documentation you can generate is the documentation that does not exist yet. If your project has no README, start there. If new developers are lost, write the architecture guide. If your API is undocumented, generate the API reference. Each document you add reduces friction for everyone who works with your code.

With the core workflows covered — generation, review, debugging, and documentation — we're ready to move into Part III and explore Claude Code's more advanced capabilities.

Part III: Project Management and Configuration

"Programs must be written for people to read, and only incidentally for machines to execute."

— Harold Abelson

Chapter 8: The CLAUDE.md System

Every developer who has worked on a team knows the experience: you join a new project, clone the repository, and spend the first two days figuring out unwritten rules. Tabs or spaces? Where do new components go? Preferred testing framework? Commit message format? These conventions live in the heads of senior developers, scattered across old Slack messages and forgotten wiki pages.

CLAUDE.md changes that. It is a simple Markdown file that gives Claude Code persistent, project-specific instructions — the context that makes the difference between generated code that feels foreign and code that looks like your team wrote it. If you take one thing from this book and apply it to every project you touch, make it this: write a CLAUDE.md file.

What CLAUDE.md Is

At its core, CLAUDE.md is just a Markdown file. No special syntax, no schema to validate against, no required sections. Claude Code reads it automatically whenever it starts a session in a directory that contains one (or a parent directory, as we'll see). The contents become part of Claude Code's context — background instructions that shape every response and action it takes during that session.

Think of CLAUDE.md as a conversation you have once instead of repeating every time. Without it, you find yourself typing variations of the same instructions session after session: "We use TypeScript strict mode in this project," "Tests go in a `__tests__` directory next to the source file," "We follow the Airbnb style guide." With a CLAUDE.md file, you say it once and Claude Code remembers — for every session, every developer on your team, every future collaborator who clones the repo.

The Three Levels

CLAUDE.md operates at three levels, and understanding how they interact is essential to using the system effectively.

Project-Level CLAUDE.md

This is the most common and most important level. You place a `CLAUDE.md` file in the root of your project repository, right alongside your `package.json`, `Cargo.toml`, or `pyproject.toml`. Every developer who clones the repo and runs Claude Code in it gets these instructions automatically.

Project-level CLAUDE.md should contain everything specific to this codebase: architecture decisions, naming conventions, preferred libraries, testing patterns, build commands, and anything else a new developer would need to know.

User-Level CLAUDE.md

You can place a CLAUDE.md file at `~/.claude/CLAUDE.md` (your home directory's `.claude` folder). These instructions apply to every Claude Code session you run, regardless of which project you are in. This is the place for your personal preferences — preferred code style when no project convention exists, favorite testing patterns, or instructions about how you like explanations formatted.

User-level instructions have lower priority than project-level ones. If your personal CLAUDE.md says "use tabs" but the project's CLAUDE.md says "use spaces," the project wins. This is the right behavior — project conventions should override personal preference.

Directory-Level CLAUDE.md

You can place CLAUDE.md files in subdirectories of your project. These apply when Claude Code works with files in that directory or its children. Particularly useful in monorepos where different packages have different conventions, or in projects where the frontend and backend follow different patterns.

For example, in a monorepo with a React frontend and a Go backend, you might have:

```
my-monorepo/
  CLAUDE.md                    # Shared project conventions
  packages/
    frontend/
      CLAUDE.md                # React-specific conventions
```

```
  backend/
    CLAUDE.md                    # Go-specific conventions
```

When Claude Code works on a file in `packages/frontend/`, it picks up the root CLAUDE.md and the frontend-specific one. When it works on backend code, it gets the root and the backend instructions instead.

Writing Effective CLAUDE.md Files

A CLAUDE.md file can contain anything, but not everything is equally useful. After working with hundreds of projects, clear patterns have emerged about what makes a CLAUDE.md file helpful versus just noise.

Start with Architecture

The single most valuable thing you can put in a CLAUDE.md file is a concise description of your project's architecture. Not a novel — a few paragraphs that describe the major components, how they interact, and where the important code lives.

```
### Architecture

This is a web application with a React frontend and Express backend.

- `src/client/` -- React app (TypeScript, Vite, React Router)
- `src/server/` -- Express API server (TypeScript, Prisma ORM)
- `src/shared/` -- Shared types and utilities used by both client and server
- `prisma/` -- Database schema and migrations

The client communicates with the server via REST API calls defined in
`src/client/api/`. The server uses Prisma to interact with a PostgreSQL database.
```

This kind of overview costs five minutes to write and saves Claude Code from having to piece together your project structure by reading dozens of files. It orients every session in the right direction from the start.

Declare Your Conventions

Explicit conventions eliminate ambiguity. Be specific about patterns you care about:

```
### Conventions
```

```
- All new files must be TypeScript (.ts or .tsx), never JavaScript
- Use named exports, not default exports
- React components use functional components with hooks, never class components
- CSS uses Tailwind utility classes; no custom CSS files except for global style
s
- All API routes follow RESTful naming: GET /users, POST /users, GET /users/:id
- Error responses use the format: { error: { code: string, message: string } }
- Database queries go through the repository pattern (see src/server/repositorie
s/)
```

Notice how specific these are. "Use TypeScript" is good. "Use functional components with hooks, never class components" is better — it tells Claude Code not just what to do but what to avoid.

Specify Your Commands

Claude Code can run commands, but it doesn't always know which commands your project uses. Tell it:

```
### Commands

- `npm run dev` -- Start development server (client and server concurrently)
- `npm run test` -- Run all tests with Vitest
- `npm run test:watch` -- Run tests in watch mode
- `npm run lint` -- Run ESLint
- `npm run typecheck` -- Run TypeScript compiler in check mode
- `npm run db:migrate` -- Run pending Prisma migrations
- `npm run db:seed` -- Seed the database with test data
```

Especially valuable when your project has non-obvious commands. If your test command is `make test-unit` or `./gradlew test --tests "com.example.*"`, Claude Code won't guess that on its own.

Include Patterns and Anti-Patterns

Sometimes the most useful instruction is "do not do this":

```
### Anti-Patterns (Do Not Use)

- Do not use `any` type in TypeScript -- use `unknown` and narrow with type guar
ds
- Do not use `console.log` for debugging -- use the project logger (`src/shared/
logger.ts`)
- Do not write raw SQL queries -- always use Prisma's query builder
```

```
- Do not import from parent directories in the client code -- use path aliases
(@/components, @/hooks, etc.)
```

Anti-patterns encode hard-won lessons. Maybe you spent a week debugging an issue caused by raw SQL queries bypassing Prisma's type safety. Encoding that lesson in CLAUDE.md means Claude Code won't make that mistake.

Keep It Current

Here is the most important advice about CLAUDE.md files: treat them as living documentation. They should evolve with your project. When you adopt a new pattern, add it. When you deprecate an old convention, update the file. When you notice Claude Code keeps making a particular mistake, add a note about it.

Some teams add updating CLAUDE.md to their definition of done for architectural decisions. Made a decision to switch from REST to GraphQL? Update CLAUDE.md. Adopted a new testing library? Update CLAUDE.md. This discipline pays enormous dividends over time.

What Not to Put in CLAUDE.md

CLAUDE.md is not the right place for everything. Avoid these common mistakes:

Do not duplicate your README. Your CLAUDE.md is instructions for an AI coding assistant, not documentation for human developers. There is overlap, but the audiences are different. Claude Code doesn't need your installation instructions or your contribution guide — it needs to know how to write code that fits your project.

Do not include large code blocks. Every token in CLAUDE.md counts against your context window. A few short examples are fine, but don't paste entire files or long configuration blocks. Reference where things live instead: "See `src/server/middleware/auth.ts` for the authentication pattern to follow."

Do not be vague. "Write good code" is useless. "Write clean code with descriptive variable names" is only slightly better. "Use camelCase for variables, PascalCase for types and components, UPPER_SNAKE_CASE for constants" is useful.

Do not include instructions that change frequently. If something changes every sprint, it probably doesn't belong in CLAUDE.md. The file should contain stable conventions, not temporary workarounds.

Examples for Different Project Types

Different kinds of projects benefit from different CLAUDE.md contents. Here are templates you can adapt.

Web Application (React + Node.js)

```
## Project: Acme Dashboard

### Overview
Internal dashboard for Acme Corp. React frontend with Node.js/Express backend.
PostgreSQL database with Prisma ORM. Authentication via Auth0.

### Architecture
- `src/client/` -- React 18 with TypeScript, Vite, React Router v6
- `src/server/` -- Express with TypeScript, organized by feature modules
- `src/shared/` -- Shared types, constants, and validation schemas (Zod)
- `prisma/` -- Schema and migrations

### Conventions
- TypeScript strict mode everywhere
- React: functional components, hooks, no class components
- State management: React Query for server state, Zustand for client state
- Styling: Tailwind CSS, no custom CSS unless absolutely necessary
- API: RESTful routes, Zod validation on all inputs
- Tests: Vitest for unit tests, Playwright for E2E
- All dates stored as UTC in the database, converted to local time in the UI

### File Naming
- React components: PascalCase (UserProfile.tsx)
- Utilities and hooks: camelCase (useAuth.ts, formatDate.ts)
- Test files: adjacent to source, same name with .test suffix (UserProfile.test.
tsx)

### Common Tasks
- `npm run dev` -- Start dev server
- `npm test` -- Run tests
- `npm run lint` -- Lint and typecheck
```

CLI Tool (Python)

```
## Project: datapipe

### Overview
CLI tool for data pipeline management. Lets users define, run, and monitor
ETL pipelines from the command line.

### Architecture
- `src/datapipe/` -- Main package
  - `cli/` -- Click-based CLI definitions
  - `core/` -- Pipeline engine, scheduler, executor
  - `connectors/` -- Database and API connectors
  - `models/` -- Pydantic models for pipeline configuration
- `tests/` -- Test suite (pytest)

### Conventions
- Python 3.12+, type hints on all functions
- Use Pydantic for data validation and serialization
- Click for CLI interface, not argparse
- Logging via structlog, not print statements
- All connectors implement the BaseConnector protocol in `connectors/base.py`
- Configuration files are TOML format

### Commands
- `make test` -- Run pytest with coverage
- `make lint` -- Run ruff and mypy
- `make format` -- Run ruff format
- `pip install -e ".[dev]"` -- Install in development mode
```

Monorepo (Multiple Packages)

Root-level CLAUDE.md:

```
## Project: Orion Platform

### Overview
Monorepo containing the Orion Platform services and shared libraries.
Uses Turborepo for build orchestration and pnpm for package management.

### Structure
- `apps/web/` -- Next.js customer-facing application
- `apps/admin/` -- Next.js admin panel
- `apps/api/` -- Fastify API server
- `packages/ui/` -- Shared React component library
```

```
- `packages/db/` -- Prisma client and database utilities
- `packages/config/` -- Shared ESLint, TypeScript, and Tailwind configs

### Conventions
- pnpm for package management (never npm or yarn)
- Shared dependencies go in the root package.json
- Package-specific dependencies go in the package's own package.json
- All packages use TypeScript strict mode
- Inter-package imports use the workspace protocol: `"@orion/ui": "workspace:*"`

### Commands
- `pnpm dev` -- Start all apps in development mode
- `pnpm build` -- Build all packages and apps
- `pnpm test` -- Run all tests
- `pnpm lint` -- Lint everything
- `turbo run build --filter=apps/web` -- Build only the web app and its dependencies
```

Library / SDK

```
## Project: httpcache

### Overview
A lightweight HTTP caching library for Go. Provides middleware for
net/http that adds RFC 7234 compliant caching.

### Architecture
- `cache.go` -- Core cache implementation
- `middleware.go` -- HTTP middleware
- `policy.go` -- Cache policy evaluation (RFC 7234)
- `store/` -- Storage backends (memory, redis, filesystem)
- `internal/` -- Internal utilities, not part of the public API

### Conventions
- Follow Go standard library conventions
- All exported types and functions must have doc comments
- Error types implement the `error` interface and include context
- Table-driven tests with descriptive subtest names
- No external dependencies in the core package; storage backends may
  use driver libraries (e.g., go-redis)
- Maintain backward compatibility: no breaking changes to exported API

### Commands
- `go test ./...` -- Run all tests
- `go test -race ./...` -- Run tests with race detector
```

```
- `go vet ./...` -- Run static analysis
- `golangci-lint run` -- Run comprehensive linting
```

CLAUDE.md as Living Documentation

One of the most powerful aspects of CLAUDE.md is that it functions as living, working documentation. Unlike a wiki page that gets written once and slowly rots, your CLAUDE.md is used every day. When it is wrong, you notice immediately — Claude Code starts generating code that doesn't match your current patterns. This creates a natural feedback loop that keeps the documentation accurate.

Some teams review CLAUDE.md changes in pull requests, just like code changes. This is smart. If someone updates the CLAUDE.md to say "use Zustand for state management" and the team hasn't agreed to that, the PR review catches it.

You can also use CLAUDE.md to document architectural decisions in-line, providing a lightweight alternative to formal Architecture Decision Records:

```
### Decisions

- We use server-side rendering for the marketing pages but client-side rendering
  for the application. This is because the marketing pages need SEO and the
  application needs real-time interactivity.

- We chose Prisma over raw SQL or Drizzle because the team values type safety
  and auto-generated migrations more than raw query performance.

- Authentication uses session-based auth (not JWT) because our security team
  requires server-side session revocation capability.
```

These notes are gold. They tell Claude Code not just what your conventions are but why. When it hits a situation where two patterns might apply, understanding the reasoning helps it make the right choice.

Advanced Patterns

As you get more comfortable with CLAUDE.md, you can use it for more sophisticated purposes.

Conditional Instructions

You can write instructions that apply only in certain contexts:

```
### When Writing Tests
- Always test the happy path and at least two error cases
- Use factory functions from `tests/factories/` for test data
- Mock external API calls using MSW (Mock Service Worker)
- Never make real HTTP requests in unit tests

### When Writing Migrations
- Always include a rollback (down migration)
- Test migrations against a copy of production data before merging
- Add a comment explaining why the migration is needed
```

Referencing Other Files

Keep your CLAUDE.md concise by pointing to detailed documentation elsewhere:

```
### API Design
Follow the API design guide in `docs/api-design.md`. Key points:
- All endpoints return JSON
- Use HTTP status codes correctly
- Include pagination for list endpoints
```

Team Onboarding Context

For projects with complex domain knowledge, a brief domain overview is remarkably effective:

```
### Domain Context
This is a healthcare scheduling application. Key domain terms:
- Provider: a doctor, nurse practitioner, or physician assistant
- Encounter: a scheduled appointment between a provider and a patient
- Slot: an available time window for scheduling an encounter
- Block: a time range when a provider is unavailable (vacation, meeting, etc.)
```

This kind of context means Claude Code uses the right terminology in variable names, comments, and log messages — making the generated code feel natural to domain experts on your team.

Custom Slash Commands

Beyond CLAUDE.md files, you can create project-specific slash commands by placing Markdown files in a `.claude/commands/` directory in your project. Each file becomes a command that any developer on the project can invoke during a Claude Code session.

For example, create a file at `.claude/commands/deploy-check.md` with the contents:

```
Run the test suite, check for TypeScript errors, and verify the build completes
successfully. Report any issues found.
```

Now any developer on the project can type `/deploy-check` during a Claude Code session, and Claude Code executes those instructions. The file name (minus the `.md` extension) becomes the command name. If you nest commands in subdirectories — for example `.claude/commands/project/deploy-check.md` — the subdirectory name becomes a prefix, invoked as `/project:deploy-check`.

A lightweight way to standardize common workflows across your team — pre-commit checks, code review checklists, migration procedures, or any multi-step task you find yourself describing repeatedly. The commands live in your repository, so they evolve with the project and are available to everyone who clones it.

Getting Started

If you don't have a CLAUDE.md file in your project yet, here is a recipe for creating one. Start a Claude Code session and try this prompt:

> Explore this codebase and suggest what should go in a CLAUDE.md file. Look at the project structure, existing patterns, testing setup, and any configuration files. Give me a draft CLAUDE.md.

Claude Code will analyze your project and produce a reasonable starting draft. Review it, refine it, and commit it to your repository. Then iterate: every time Claude Code does something that doesn't match your conventions, update the CLAUDE.md and try again. Within a week, you'll have a CLAUDE.md file that makes Claude Code feel like a team member who has been on the project for months.

Your CLAUDE.md file is the highest-leverage artifact you can create for AI-assisted development. Invest the time to write a good one, keep it updated, and watch the quality of every Claude Code interaction improve.

Chapter 9: Working with Large Codebases

The demo projects are easy. A todo app with twelve files. A portfolio site with a handful of components. Claude Code handles these effortlessly — reads the entire project, understands everything, produces clean output. Then you point it at your actual codebase: 2,000 files, 400,000 lines of code, a dependency graph that looks like a plate of spaghetti. Can Claude Code still help?

Yes. But you need to understand how it works under the hood, and you need strategies for guiding it through large projects effectively. This chapter covers both.

How Claude Code Handles Context

Every interaction with Claude Code is bounded by a context window — the maximum amount of text the model can consider at one time. The window is large (and growing with each model generation), but not infinite. In a codebase with hundreds of thousands of lines, you cannot fit everything into context simultaneously.

Claude Code manages this constraint through intelligent selection. When you ask a question or give it a task, it doesn't blindly try to read every file. Instead, it uses a combination of strategies:

File search and indexing. Claude Code can search your codebase by file name, by content (using grep-like patterns), and by structural elements like function names, class names, and imports. It reads the files that are relevant to your current task, not the entire project.

Incremental exploration. When Claude Code needs to understand how something works, it follows the trail: reading a file, noticing an import, reading the imported file, and so on. This mirrors how a human developer navigates unfamiliar code — you don't read every file in order; you follow the dependency chain from your starting point.

Conversation memory. Within a session, Claude Code remembers everything it has read and everything you have discussed. If it read your authentication middleware twenty messages ago, it still knows what is in there. This cumulative context builds up a working understanding of the relevant parts of your codebase.

CLAUDE.md context. Your CLAUDE.md file gives Claude Code a head start by providing architecture overviews, file structure descriptions, and conventions from the very first message (see Chapter 8).

The practical implication: Claude Code doesn't need to understand your entire codebase to be useful. It needs to understand the parts relevant to your current task. Your job is to help it find and focus on the right parts.

Orienting in Unfamiliar Code

One of the most underappreciated uses of Claude Code is as an exploration tool. When you're dropped into a codebase you've never seen before — a new job, an inherited project, an open-source library you need to contribute to — Claude Code can orient you faster than any other tool.

The Exploration Prompt

Start broad and narrow down. Here is a sequence that works well for getting Claude Code oriented in a new codebase:

Start with the big picture:

> Explore this codebase and give me a high-level overview. What does this project do, what technologies does it use, and how is it structured?

Claude Code reads your package files, configuration files, directory structure, and a sampling of source files to build a picture. The result is usually a solid orientation in under a minute.

Then drill into specifics:

> Walk me through how a request flows from the API endpoint to the database and back. Use a specific example — pick a simple CRUD endpoint.

This forces Claude Code to trace through multiple files: the route handler, any middleware, the controller or service layer, the data access layer, and the response formatting. You learn the architecture by following a real path through it.

Then explore the parts you'll work on:

> I need to add a new feature to the reporting module. Show me the existing reporting code and explain how reports are currently generated.

Now Claude Code focuses on the area you care about, and it already has the broader context from your earlier questions.

Reading Code Strategically

When working with large codebases, resist the urge to have Claude Code read everything. Instead, be strategic about what you ask it to look at.

Start with entry points. Every application has entry points: the main function, the route definitions, the event handlers. These tell you what the application does and where the important code paths begin.

Follow the data. If you're trying to understand a feature, follow the data through the system. Where does user input enter? How is it validated? Where is it stored? How is it retrieved and displayed? Each step leads you to the relevant files.

Read tests for behavior. Test files are often the best documentation of how code is supposed to work. Ask Claude Code to read the test files for a module you're trying to understand — they show you the expected inputs, outputs, and edge cases.

Check the configuration. Configuration files often reveal critical architecture details: database connections, API integrations, feature flags, environment-specific behavior. A quick look at the config explains a lot of seemingly odd code.

Working with Monorepos

Monorepos present a specific challenge: multiple projects, sometimes in different languages, with different conventions and tooling. Claude Code handles monorepos well, but you need to be thoughtful about how you direct its attention.

Scoping Your Work

When you're working on a specific package within a monorepo, tell Claude Code explicitly:

> I am working in the packages/auth-service directory. This is a Fastify-based authentication service. For the next set of tasks, focus on this package.

This framing helps Claude Code avoid reading unrelated packages and keeps the context focused on what matters.

Cross-Package Work

Sometimes you need to make changes that span multiple packages — updating a shared library and all its consumers, for example. Be explicit about the scope:

> I need to add a new field to the User type in packages/shared-types and then update all packages that import User to handle the new field. Let's start by finding everywhere User is imported.

Claude Code searches across the monorepo, builds a list of affected files, and works through the changes systematically. One of its strongest capabilities — coordinated changes across many files that a human developer would need hours to track down and update.

Directory-Level CLAUDE.md

You can place CLAUDE.md files in subdirectories (see Chapter 8 for details on the three-level system). In a monorepo, this is almost mandatory. Each package should have its own CLAUDE.md with package-specific conventions, commands, and architecture notes. The root CLAUDE.md covers shared conventions and the overall structure.

The /compact Command

Long sessions with Claude Code are productive, but they accumulate context. Every file read, every response generated, every code block discussed — it all stays in the conversation history. Eventually, you'll notice responses slow down or Claude Code bumps against its context limit.

The `/compact` command is your solution. When you run it, Claude Code summarizes the current conversation into a compressed form, discarding the raw details while preserving the key information: what files have been discussed, what decisions have been made, what the current task is, and what has been accomplished.

When to Use /compact

After completing a major subtask. If you've spent twenty messages setting up a database schema and are now moving on to the API layer, compact first. The schema details are in the files now; Claude Code can re-read them if needed.

When you notice slowdown. If Claude Code's responses are taking noticeably longer, the context may be getting large. Compacting improves response speed.

Before switching topics. If you've been debugging an authentication issue and want to switch to the reporting feature, compacting clears out the debugging context while preserving the session's general understanding of the project.

During long implementation sessions. If you're working through a large feature that takes fifty or more messages, compact every fifteen to twenty messages. Think of it as saving your progress.

What /compact Preserves

The summary /compact creates is not a truncation — it is an intelligent distillation. It preserves:

- The project structure and architecture you've discussed
- Key decisions made during the session
- Files that have been modified and why
- The current task and its status
- Important context about the codebase that surfaced

What it discards is the raw back-and-forth: full contents of files that were read, intermediate versions of code that were iterated on, detailed explanations that have already served their purpose.

What /compact Does Not Replace

Compacting is not a substitute for starting a new session. If you've finished one task and are starting something entirely different, a fresh session gives you a clean slate with full context available. Use compacting for continuity within a single broader task; use new sessions for truly separate work.

Session Management

Knowing when to continue a session and when to start fresh is a skill that separates novice Claude Code users from experienced ones.

Continue When

- You're still working on the same feature or task
- The context Claude Code has built up is valuable and would be tedious to reconstruct
- You're iterating on code Claude Code wrote earlier in the session
- You need to reference decisions made earlier in the conversation

Start Fresh When

- You're switching to a completely different task
- The previous session went off track and you want a clean start
- You've made significant changes to the codebase outside of Claude Code (manual edits, merging branches, pulling updates) and want Claude Code to see the current state fresh
- The session has been running a long time and feels sluggish even after compacting

The Handoff Pattern

When you're ending a session but plan to continue the work later, ask Claude Code to write a handoff summary:

> I am going to end this session. Write a summary of where we are, what has been done, what remains, and any important context for the next session. Format it so I can paste it as the first message in a new session.

Claude Code produces a concise summary you can save and use to bootstrap your next session. Especially useful for multi-day features or when handing off work to a teammate.

Some developers keep a running notes file — `WORKNOTES.md` or `.claude/session-notes.md` — where they paste these summaries. It becomes a log of the development process useful for more than just Claude Code.

Performance Considerations

Large codebases can stress any tool, and Claude Code is no exception. Here are strategies for keeping things fast and responsive.

Be Specific in Your Prompts

Vague prompts force Claude Code to do broad searches. Instead of:

> Find the bug in the user module.

Try:

> There is a bug where updating a user's email does not trigger the email verification flow. The update endpoint is in src/api/users.ts and the verification logic is in src/services/email-verification.ts. Find why the verification is not being triggered.

The specific prompt sends Claude Code directly to the relevant files instead of searching through the entire user module.

Avoid Unnecessary File Reads

If you know a particular file is irrelevant to your current task, you don't need to mention it. Claude Code won't read files unless it has a reason to. But if your prompt is ambiguous — "look at the configuration" when there are forty configuration files — Claude Code may read more than necessary.

Use CLAUDE.md to Reduce Exploration

A good CLAUDE.md file with an architecture overview cuts the exploration Claude Code needs to do (see Chapter 8 for how to write one). If CLAUDE.md says the email verification logic is in `src/services/email-verification.ts`, Claude Code goes there directly instead of searching.

Manage Your Context Budget

Every file Claude Code reads consumes context. In a long session, you may find that Claude Code has read so many files it is running low on room for its actual responses. The /compact command helps, but so does being intentional about what you ask Claude Code to read.

A useful mental model: context is a budget. Every file read, every long explanation, every code block generated is an expenditure. Spend it on what matters for your current task.

Working with Generated Code

Large codebases often include generated code — Protocol Buffer definitions, GraphQL type generators, OpenAPI client code, and so on. These files can be large and are usually not useful for Claude Code to read directly. Add a note to your CLAUDE.md:

```
### Generated Code (Do Not Modify)
- `src/generated/` -- Auto-generated GraphQL types (run `npm run codegen` to reg
enerate)
- `src/proto/` -- Generated Protocol Buffer code
These files should never be edited directly.
```

This tells Claude Code to focus on the source definitions rather than the generated output.

Strategies for Massive Projects

Some codebases are enormous — millions of lines, thousands of files, decades of accumulated code. These require extra discipline.

Divide and Conquer

Break large tasks into focused sub-tasks. Instead of "refactor the authentication system," start with "refactor the password hashing in `src/auth/password.ts`" and work outward from there. Each sub-task can be a focused Claude Code session with a manageable scope.

Create a Map

For very large projects, consider maintaining a higher-level architecture document your CLAUDE.md references. A document that maps major features to their primary files and directories is enormously valuable:

```
### Feature Map
- User authentication: src/auth/, src/middleware/auth.ts, src/models/user.ts
- Payment processing: src/billing/, src/services/stripe.ts
- Notification system: src/notifications/, src/jobs/email-worker.ts
- Reporting: src/reports/, src/services/analytics.ts
```

Use Multiple Sessions

For changes that span many areas of a large codebase, use multiple focused sessions rather than one massive session. Do the database changes in one session, the API changes in another, and the frontend changes in a third. Each session stays focused and has plenty of context for its area.

Trust but Verify

In large codebases, Claude Code might not find every place that needs to change. After it makes changes, run your tests, your linter, and your type checker. These tools catch anything Claude Code missed — a forgotten import update, a type mismatch, a broken test. Claude Code is an excellent first pass; your existing tooling is the safety net.

Using Git Worktrees for Parallel Sessions

Git worktrees let you check out multiple branches of the same repository into separate directories simultaneously. This pairs exceptionally well with Claude Code: you can run independent sessions in each worktree — working on different features, comparing approaches, or testing a fix on one branch while continuing development on another.

To create a worktree:

```
git worktree add ../my-project-feature-x feature-x
```

This creates a new directory at `../my-project-feature-x` with the `feature-x` branch checked out. Add `-b feature-x` after `add` to create the branch at the same time if it does not already exist. Open a separate terminal, navigate to that directory, and run `claude` there. Each session operates on its own branch with its own working state. No interference between sessions because each worktree is a fully independent checkout.

Particularly useful for large codebases where switching branches is expensive (triggering rebuilds, re-installing dependencies, or resetting local state). Instead of switching, you keep multiple branches alive in parallel. When you're done, clean up with `git worktree remove ../my-project-feature-x`.

Resuming Previous Sessions

Large codebase work often spans multiple sittings. Claude Code provides two flags for picking up where you left off:

`--continue` (or `-c`) resumes the most recent Claude Code session in the current directory. This is the quick option — you close your terminal, come back later, and run `claude --continue` to pick up the same conversation with all its accumulated context.

`--resume` (or `-r`) lets you choose from a list of recent sessions or resume a specific session by ID. Run `claude --resume` without an argument to see your recent sessions, or `claude --resume <session-id>` to jump directly to a specific one.

Both flags restore the conversation history so Claude Code remembers the files it read, the decisions you made, and the current state of the task. Essential for ongoing work on large codebases where rebuilding context from scratch would waste significant time and tokens. The handoff pattern described in the Session Management section above is still valuable for multi-day work, but `--continue` and `--resume` make single-day interruptions seamless.

Putting It All Together

Working with large codebases is where Claude Code proves its value. A human developer might spend a day tracing through unfamiliar code to understand how a feature works. Claude Code does it in minutes. A team might spend a week on a cross-cutting refactor, manually updating hundreds of files. Claude Code handles the mechanical parts in a fraction of the time, freeing you to focus on the design decisions.

The pattern is the same throughout. You bring the architecture knowledge and the judgment calls. Claude Code brings speed and tireless thoroughness. Codebases that would take weeks to absorb solo become tractable in days.

In the next chapter, we'll put this into practice with multi-file operations — the kind of coordinated, cross-cutting changes large codebases demand and Claude Code handles exceptionally well.

Chapter 10: Multi-File Operations

Most real features don't live in a single file. Adding a new API endpoint means creating a route handler, a controller, a service function, a database migration, types, validation schemas, tests, and documentation. Renaming a concept means updating every file that references it. Adding logging means touching dozens of files across every layer of your application.

This is where Claude Code pulls ahead of every other AI coding tool. While autocomplete tools help you write one file at a time, Claude Code sees the whole picture and makes coordinated changes across your entire codebase. This chapter teaches you how to use that capability effectively.

The Mental Model for Multi-File Work

When you ask Claude Code to make a change that spans multiple files, it follows a pattern that mirrors an experienced developer:

1. **Understand the scope.** Claude Code figures out which files need to change. It searches the codebase, follows imports, reads related files, and builds a mental map of the change.

2. **Plan the changes.** Before writing any code, Claude Code develops a plan — which files to modify, which to create, and in what order. You'll often see it explain this plan before executing.

3. **Execute systematically.** Claude Code makes the changes file by file, maintaining consistency across all of them. Types match, imports align, naming conventions hold.

4. **Verify.** After making the changes, Claude Code can run your tests, linter, or type checker to confirm everything fits together.

You can participate in any of these steps. Ask Claude Code to show you the plan before it executes. Redirect it if the plan misses something. Ask it to verify specific aspects of the change. The more complex the change, the more valuable this collaboration becomes.

Creating Features That Touch Many Files

The most common multi-file operation is creating a new feature. Let's walk through how this works in practice.

Say you're building an e-commerce application and need to add a wishlist feature. A good prompt sets the scope:

> Add a wishlist feature to this application. Users should be able to add products to their wishlist, remove products, and view their wishlist. Follow the existing patterns in the codebase for how features are structured.

That last sentence — "follow the existing patterns" — is crucial. Claude Code examines your existing features (the shopping cart, the order history, the product catalog) and replicates their structure for the new feature. If your cart feature has a route file, a controller, a service, a repository, types, and tests, the wishlist feature gets the same treatment.

Claude Code typically creates:

- A database migration adding the wishlist table
- A Prisma model or ORM entity for wishlists
- A repository or data access layer
- A service layer with business logic
- API route handlers
- Request/response type definitions
- Validation schemas
- Unit tests for the service layer
- Integration tests for the API endpoints

All of these files are internally consistent: types used in the route handlers match the types in the service layer, which match the database schema. Imports are correct. Naming follows your project's conventions.

Incremental Feature Building

For complex features, you don't have to do everything in one prompt. Build incrementally:

> Let's add a wishlist feature. Start with just the database schema and the data access layer. I want to review that before we build the API on top of it.

Review the output, request any changes, then continue:

> That looks good. Now add the API endpoints for CRUD operations on the wishlist, following the same patterns as the cart endpoints.

Then:

> Now add the frontend components. Look at how the cart page is built and follow the same patterns for the wishlist page.

The incremental approach gives you control over the architecture at each layer while still leveraging Claude Code's ability to generate consistent, pattern-matching code.

Coordinated Changes Across a Codebase

Sometimes you're not adding something new — you're changing something that already exists everywhere. These coordinated changes are tedious for humans but natural for Claude Code.

Renaming a Concept

Renaming is deceptively complex. Changing a variable name in one file is trivial. Renaming a domain concept across an entire codebase — database columns, API fields, variable names, type definitions, test fixtures, documentation, log messages — is its own project.

> Rename the concept "Organization" to "Workspace" throughout the codebase. This includes database tables and columns, API endpoints, type definitions, variable names, file names, and test fixtures. Do not change the actual database table names yet — just the application code. We will handle the migration separately.

Claude Code will:

- Find every file that references "Organization" or "organization"

- Update type names (`Organization` to `Workspace`, `OrganizationMember` to `WorkspaceMember`)
- Update variable names (`org`, `organization` to `workspace`, `ws`)
- Update API routes (`/organizations` to `/workspaces`)
- Update test descriptions and fixture names
- Update comments and log messages
- Preserve the database column names as instructed

A change like this might touch fifty or a hundred files. Claude Code handles the mechanical work while you review the results and handle the edge cases.

Updating an API Contract

When you change an API response format, you need to update the server-side code, the client-side code, the types, the tests, and possibly the documentation. Claude Code excels at these cascading changes:

> Change the API error response format from { message: string, code: number } to { error: { message: string, code: string, details?: object } }. Update the error handling middleware, all places where errors are constructed, the frontend error handling, the types, and the tests.

Dependency Upgrades with Breaking Changes

When a major dependency upgrade introduces breaking changes, Claude Code can handle the migration:

> We are upgrading React Router from v5 to v6. Update all route definitions, replace useHistory with useNavigate, replace Switch with Routes, and update all component files that use routing.

Claude Code searches for every usage of the old API, understands the migration path, and applies the changes consistently. It knows common library migration patterns and handles the translation from old API to new API accurately.

Intelligent Find-and-Replace

Claude Code's search-and-replace capability goes far beyond simple text substitution. It understands code structure, so it can make changes that a regex could never handle.

Structural Changes

Consider this task:

> Convert all of our API handler functions from the callback pattern to async/await. They currently look like: `function handler(req, res, next) { service.getData().then(data => res.json(data)).catch(next) }`. Convert them to: `async function handler(req, res, next) { try { const data = await service.getData(); res.json(data); } catch (err) { next(err); } }`

A regex cannot do this reliably — callback patterns vary, variable names differ, some handlers have multiple chained promises. But Claude Code understands the code structure. It reads each handler, understands what it does, and rewrites it using async/await while preserving the logic.

Pattern-Aware Replacement

> Every component that uses the old `withAuth` higher-order component should be refactored to use the `useAuth` hook instead. Remove the HOC wrapper and add the hook call at the top of the component function.

Claude Code understands what the HOC does, how the hook replaces it, and how to restructure each component. Each component is different — different props, different usage of the auth data — and Claude Code handles them individually rather than applying a blind pattern.

Adding Cross-Cutting Concerns

Cross-cutting concerns — logging, error handling, authentication, metrics, caching — touch every part of your application. Adding them retroactively is one of the most tedious tasks in software development. It's also one of Claude Code's specialties.

Adding Structured Logging

> Add structured logging to all API endpoint handlers. Use the winston logger imported from src/lib/logger. Log the request method, path, and duration for every request. Log errors with the full error object. Follow the pattern established in src/api/health.ts.

Claude Code reads the reference file, understands the pattern, and applies it to every endpoint handler. Each handler gets logging that makes sense for its specific context — a user creation endpoint logs the new user ID, a search endpoint logs the query parameters, and so on.

Adding Error Boundaries

> Add error boundaries around every page-level React component. Use the ErrorBoundary component from src/components/ErrorBoundary. Include a fallback UI that shows the page name and a retry button.

Adding Metrics

> Add Prometheus metrics to all database queries. Use the histogram from src/metrics/db.ts to track query duration. Label each metric with the operation name and table name.

In each case, Claude Code doesn't just add boilerplate. It adapts the cross-cutting concern to each specific location, using appropriate labels, messages, and context.

Database Schema Changes with Cascading Updates

Database changes are among the most far-reaching modifications you can make. A new column, a changed relationship, or a normalized table ripples through migrations, models, repositories, services, API endpoints, validation schemas, and frontend forms. Claude Code handles the cascade:

> Add a "status" field to the Order model. It should be an enum with values: pending, processing, shipped, delivered, cancelled. Default to "pending". Update the database migration, the Prisma schema, the

> Order type, the API endpoints (add filtering by status), the order service, and the frontend order list (show status with a colored badge).

Claude Code works through each layer systematically, keeping the enum values consistent everywhere, the types flowing correctly from the database to the frontend, and the new field validated on input and displayed on output.

Real-World Workflow: Adding Authentication to an Existing Application

Let's walk through a complete multi-file operation from start to finish. You have an existing Express application with a React frontend, and you need to add user authentication. The app currently has no concept of users.

Step 1: Plan the Approach

Start with planning, not coding:

> I need to add user authentication to this application. Currently there are no users — everything is anonymous. I want session-based authentication with email/password login. Before writing any code, analyze the codebase and give me a plan for what needs to change.

Claude Code explores the codebase and produces a plan:

- Database: add a users table with email, hashed password, and profile fields
- Backend: add auth routes (register, login, logout), session middleware, password hashing
- Backend: add auth middleware to protect existing endpoints
- Frontend: add login/register pages, auth context, protected route wrapper
- Frontend: update the API client to handle 401 responses
- Tests: auth-related tests at every layer

Review the plan. Maybe you want OAuth instead of email/password. Maybe JWT instead of sessions. Adjust before any code is written.

Step 2: Build the Foundation

> Let's start with the database and backend foundation. Create the user migration, the User model, and the auth utility functions (password hashing, session management). Do not add routes yet.

Claude Code creates the migration, the model, and the utility functions. Review them. Are the password hashing settings secure? Does the session configuration match your production requirements?

Step 3: Add the Auth Routes

> Now add the authentication API routes: POST /auth/register, POST /auth/login, POST /auth/logout, GET /auth/me. Include input validation and proper error handling. Follow the patterns in the existing route files.

Claude Code adds route handlers that look like they belong in your project. Validation uses whatever library your other routes use. Error handling follows your existing patterns. The response format matches your API conventions.

Step 4: Protect Existing Routes

> Add authentication middleware that checks for a valid session. Apply it to all existing API routes except the public ones (which are: GET /products, GET /products/:id, and the health check). Update each protected route handler to have access to the authenticated user.

This is where multi-file operations shine. Claude Code modifies every existing route file, adding the middleware and updating the handler signatures to include the user. If a route already does something user-specific (like showing "your recent orders"), Claude Code wires it to the authenticated user instead of whatever placeholder was there before.

Step 5: Build the Frontend

> Now add the frontend authentication. Create a login page, a registration page, an auth context provider, and a ProtectedRoute component. Update the app's routing to require authentication for all pages except the product catalog. Add a login/logout button to the header.

Claude Code creates the pages and components, wires up the auth context, wraps the router with the provider, and updates the header. Styling matches your existing components. Form validation follows the patterns used elsewhere in the frontend.

Step 6: Update the API Client

> Update the frontend API client to handle 401 responses by redirecting to the login page. Also add the session cookie handling to all API requests.

Step 7: Add Tests

> Add tests for the authentication system. Include unit tests for the password hashing utilities, integration tests for the auth API endpoints, and frontend tests for the login flow and protected routes.

Step 8: Verify Everything

> Run the full test suite and fix any failures. Then run the linter and type checker and fix any issues.

Claude Code runs your tests, finds the failures (perhaps an existing test that now needs authentication in its setup), and fixes them.

The Result

After this workflow, you have a complete authentication system that spans:

- 1 database migration
- 3–4 new backend files (routes, middleware, utilities)
- Modifications to every existing route file
- 4–5 new frontend files (pages, context, protected route component)
- Modifications to the API client and router
- 15–20 new test files and modifications to existing tests

Easily a week of manual work for a single developer. With Claude Code, the implementation takes an afternoon, and you spend most of that time reviewing and refining rather than writing boilerplate.

Tips for Multi-File Success

Be explicit about patterns to follow. "Follow the existing patterns" is one of the most powerful things you can say to Claude Code. It examines your codebase and replicates your conventions. But if you have specific patterns in mind, name them: "Follow the pattern in `src/api/orders.ts`."

Review changes in layers. Don't try to review fifty file changes at once. Build in layers: database, then backend, then frontend. Review each layer before moving to the next.

Let Claude Code run your tools. After making changes, ask Claude Code to run your tests, linter, and type checker. These tools catch inconsistencies that are hard to spot in a code review.

Commit between layers. After each successful layer, commit the changes. That gives you a clean rollback point if something goes wrong in a later layer.

Keep your CLAUDE.md current. Multi-file operations are where CLAUDE.md pays off most. If your CLAUDE.md accurately describes your project structure and conventions, Claude Code's multi-file changes will be consistent and correct from the start.

The ability to make coordinated changes across many files is not just a convenience — it changes how you think about software development. Tasks you used to avoid because of the tedium ("we should rename that concept, but it's everywhere") become practical. Improvements that would have taken a sprint become afternoon tasks. And the resulting code is consistent, because it was all generated with the same context and conventions.

Chapter 11: Testing Strategies

There is a dirty secret in software development: most codebases are undertested. Developers know they should write tests. They agree tests are valuable. But when the deadline hits, tests are the first thing that gets cut. The result is millions of lines of production code running without a safety net.

Claude Code changes the economics of testing. Writing tests is no longer the slow chore that competes with feature work for your time. Claude Code generates tests quickly, thoroughly, and in a style that matches your existing test suite. This chapter covers everything from basic unit test generation to achieving meaningful coverage on legacy codebases that have never been tested.

Generating Tests from Existing Code

The simplest and most immediately useful testing workflow: point Claude Code at existing code and ask it to write tests.

> Write unit tests for src/services/order-service.ts. Cover the happy path and edge cases for each public method. Follow the testing patterns in the existing test files.

Claude Code will:

1. Read the service file to understand its public interface
2. Read any existing test files to learn your patterns (which framework, which assertion style, how mocks are set up)
3. Read the dependencies (types, related services, repositories) to see what needs to be mocked
4. Generate comprehensive tests

The result is not a handful of trivial smoke tests. Claude Code thinks about what could go wrong:

- What happens when the input is empty, null, or malformed?
- What happens when a dependency throws an error?
- What happens at boundary conditions (zero items, maximum items, exact limit)?

- What happens with concurrent operations?
- What happens with invalid state transitions?

A typical test generation for a moderately complex service file produces twenty to forty test cases across five to ten describe blocks. Most pass immediately. The ones that fail often reveal real bugs or undocumented behavior — exactly the point of testing.

Guiding Test Generation

You can steer Claude Code toward the kind of tests you want:

> Write tests for the payment processing module. Focus on error handling — I want to make sure every failure mode is tested, including network timeouts, invalid card numbers, insufficient funds, and rate limiting from the payment provider.

Or:

> Write tests for the date utility functions. These are pure functions with no side effects, so no mocking is needed. Use parameterized tests where multiple inputs should produce similar behavior.

Or:

> Write integration tests for the /api/users endpoints. Use supertest to make actual HTTP requests against the app. Set up and tear down a test database for each test suite.

The more specific you are about what kind of tests you want, the better the result. But even a generic "write tests for this" produces thorough output.

Test-Driven Development with Claude Code

Test-driven development — writing tests before the implementation — works beautifully with Claude Code. The workflow inverts the usual pattern:

Step 1: Describe the Behavior

Start by telling Claude Code what you want to build, in terms of behavior:

> I need a function called `calculateShipping` that takes an order (with items, weights, and destination) and returns a shipping cost. Here are the rules: - Domestic orders under 2kg: $5.99 - Domestic orders 2-10kg: $12.99 - Domestic orders over 10kg: $12.99 + $1.50 per kg over 10 - International orders: 2.5x the domestic rate - Orders over $100 get free domestic shipping - Free shipping does not apply to international orders

Step 2: Generate the Tests First

> Write the tests first. Do not write the implementation yet. Test every rule listed above, plus edge cases at the boundaries.

Claude Code generates a thorough test suite:

```
describe('calculateShipping', () => {
  describe('domestic orders', () => {
    it('charges $5.99 for orders under 2kg', () => { /* ... */ });
    it('charges $12.99 for orders at exactly 2kg', () => { /* ... */ });
    it('charges $12.99 for orders between 2kg and 10kg', () => { /* ... */ });
    it('charges $12.99 for orders at exactly 10kg', () => { /* ... */ });
    it('adds $1.50 per kg over 10kg', () => { /* ... */ });
    it('provides free shipping for orders over $100', () => { /* ... */ });
    it('provides free shipping for orders at exactly $100', () => { /* ... */ });
    it('does not provide free shipping for orders at $99.99', () => { /* ... */
 });
  });

  describe('international orders', () => {
    it('charges 2.5x the domestic rate', () => { /* ... */ });
    it('does not apply free shipping for international orders over $100', () =>
{ /* ... */ });
  });

  describe('edge cases', () => {
    it('handles an order with zero items', () => { /* ... */ });
    it('handles an order with zero weight', () => { /* ... */ });
    it('throws on negative weight', () => { /* ... */ });
  });
});
```

Step 3: Write the Implementation

> Now write the implementation that makes all these tests pass.

Claude Code writes the function with all the rules correctly implemented. Every test passes.

Step 4: Refine

> Are there any edge cases the tests do not cover?

Claude Code reviews both the tests and the implementation and suggests cases you might have missed: What about Alaska and Hawaii? What about orders where some items have no weight specified? What about rounding errors on the per-kg surcharge?

The TDD workflow produces well-tested, well-specified code. The tests serve as living documentation of every business rule, and the implementation handles every specified case.

Generating Test Fixtures and Mock Data

Good tests need good data. Claude Code excels at generating realistic test fixtures that cover a range of scenarios.

Factory Functions

> Create a test factory for User objects. It should generate realistic-looking user data with sensible defaults that can be overridden. Use faker.js for realistic values.

Claude Code produces factory functions that generate diverse, realistic test data:

```
function createUser(overrides?: Partial<User>): User {
  return {
    id: faker.string.uuid(),
    email: faker.internet.email(),
    name: faker.person.fullName(),
    role: 'member',
    createdAt: faker.date.recent(),
    ...overrides,
  };
}
```

Complex Fixture Sets

For integration tests, you often need sets of related data:

> Create a test fixture that represents a realistic e-commerce scenario: 3 users, 10 products across 3 categories, 5 orders with varying statuses and items, and 8 reviews. Make the data internally consistent — reviews reference real products and users, orders contain real products with correct totals.

Claude Code generates a complete, internally consistent dataset where the foreign keys line up, the totals are mathematically correct, and the data tells a realistic story.

Mock Services

> Create a mock implementation of the PaymentGateway interface for testing. It should simulate successful charges, declined cards (when the card number starts with '4000'), network timeouts (when amount is exactly $0.01), and rate limiting (on the 11th call in a test).

Claude Code builds a mock sophisticated enough for thorough testing but simple enough to understand and maintain.

Testing Edge Cases

One of Claude Code's most valuable testing capabilities is thinking adversarially about code. Ask it to find the edge cases:

> Look at the `parseUserInput` function and write tests for every edge case you can think of. Try to break it.

Claude Code tests inputs like:

- Empty strings, whitespace-only strings, strings with only newlines
- Extremely long strings (buffer overflows, performance degradation)
- Unicode characters, emoji, right-to-left text, zero-width characters
- Special characters that might cause injection (SQL, HTML, shell)
- Null, undefined, NaN, Infinity
- Numbers at the boundaries of integer ranges
- Strings that look like numbers, booleans, or null
- Deeply nested objects if the input accepts objects
- Arrays with zero, one, and many elements

- Concurrent calls with conflicting data

You don't need to think of all these cases yourself. Claude Code has seen thousands of bugs caused by edge cases and knows where to look.

Negative Testing

> Write tests that verify our API correctly rejects invalid input. Test every endpoint with missing required fields, wrong types, values out of range, and malformed JSON.

Negative testing is tedious to write manually but critical for security and robustness. Claude Code generates it systematically, covering every field in every endpoint.

Coverage Analysis and Gap Filling

If you have a codebase with partial test coverage, Claude Code can help you identify and fill the gaps.

Finding Uncovered Code

> Run the test coverage report and identify the files and functions with the lowest coverage. List them sorted by coverage percentage, lowest first.

Claude Code runs your coverage tool, reads the report, and gives you a prioritized list of what needs tests most.

Targeted Gap Filling

> The coverage report shows that src/services/notification-service.ts has 23% coverage. The only tested function is sendEmail. Write tests for the remaining functions: scheduleNotification, cancelNotification, getNotificationHistory, and retryFailedNotifications.

Claude Code reads the uncovered functions, understands their behavior, and writes targeted tests that fill the gaps identified in the coverage report.

Meaningful vs. Meaningless Coverage

Not all coverage is equal. Claude Code understands this:

> We have 60% line coverage but I suspect a lot of it is superficial — tests that execute lines but do not meaningfully assert behavior. Review our test suite and identify tests that execute code without making meaningful assertions. Then improve them.

Claude Code finds tests that call functions but only assert they "do not throw" when they should be asserting specific return values, state changes, or side effects. It strengthens these tests with meaningful assertions.

Integration and End-to-End Tests

Unit tests verify individual functions. Integration tests verify the pieces work together. Claude Code writes both, but integration tests deserve special attention because they're harder to write well.

API Integration Tests

> Write integration tests for the order workflow: creating an order, processing payment, updating inventory, and sending confirmation. These should test the full flow through the actual service layer (not mocked), but use a test database.

Claude Code will:

- Set up the test database with seed data
- Walk through the full workflow, calling service functions in sequence
- Assert the state after each step (order in database, inventory decremented, email queued)
- Test error scenarios (payment failure, out-of-stock items)
- Clean up after each test

Database Integration Tests

> Write tests that verify our Prisma queries return the correct data with realistic scenarios. Include tests for pagination, filtering, sorting, and edge cases like empty results and exact-match boundaries.

Frontend Integration Tests

> Write Playwright tests for the checkout flow: add item to cart, go to checkout, fill in shipping information, enter payment details, submit order, verify confirmation page.

Claude Code generates end-to-end tests that drive your application the way a real user would, with proper waits, assertions, and cleanup.

Real-World Workflow: Achieving 80% Coverage on Legacy Code

Let's walk through the most common testing scenario: a legacy codebase with minimal test coverage that needs to be brought up to a reasonable standard. We'll target 80% line coverage, a common industry benchmark.

Step 1: Assess the Current State

> Run the test suite and generate a coverage report. Give me a summary: overall coverage, coverage by directory, and a list of the 20 files with the lowest coverage that have the most lines of code.

Claude Code runs your tests, reads the coverage output, and gives you a clear picture. Suppose the results show 34% overall coverage, critical services at 15%, and utilities at 60%.

Step 2: Prioritize

Not all code is equally important to test. Ask Claude Code to help prioritize:

> Given the coverage report and the codebase structure, which files should we prioritize for testing? Consider: code that handles money or payments, code that handles user authentication, code that is called frequently, and code that has had recent bug fixes.

Claude Code analyzes the codebase and produces a prioritized list. Payment service, auth middleware, and the order processing pipeline rise to the top.

Step 3: Establish Testing Patterns

Before generating tests at scale, establish your patterns:

> Look at the existing test files and identify the testing patterns in use. Then create a test utilities file at src/test/helpers.ts with common setup functions, factory functions, and custom assertions that we will use across all new tests.

This gives you a foundation that makes all subsequent tests consistent and DRY.

Step 4: Generate Tests in Priority Order

Now work through the priority list. For each module:

> Write comprehensive tests for src/services/payment-service.ts. Target at least 85% coverage of this file. Use the test helpers we just created. Include both unit tests (with mocked dependencies) and integration tests (with a test database).

Claude Code generates a thorough test suite. Run it, review it, commit it. Then move to the next file.

Step 5: Handle the Difficult Cases

Some legacy code is hard to test because it wasn't designed for testability: functions with many side effects, tight coupling between modules, global state, hidden dependencies. Claude Code can help:

> The processOrder function in src/services/order-service.ts is difficult to test because it directly calls the payment gateway, the email service, and the inventory service. Refactor it to accept these as parameters (dependency injection) so we can pass mocks in tests. Then write the tests.

A pattern you'll use repeatedly: a small refactor to make the code testable, followed by test generation. Claude Code handles both steps.

Step 6: Fill the Remaining Gaps

After covering the priority files, run coverage again:

> Run coverage again. We were at 34%. What are we at now, and what are the biggest remaining gaps?

Suppose you're at 68%. The remaining gaps are mostly utility functions, error handlers, and edge-case code paths. These are usually the easiest to test:

> Write tests for all untested utility functions in src/utils/. These are mostly pure functions, so the tests should be straightforward.

And:

> Add error path tests to the existing test suites. For each service, add tests that verify correct behavior when database queries fail, when external APIs return errors, and when input validation rejects bad data.

Step 7: Cross the Finish Line

Run coverage one more time. You're at 78%. The remaining gaps are a handful of hard-to-reach code paths:

> The coverage report shows these specific uncovered lines. For each one, write a test that exercises that code path, or explain why it cannot be reasonably tested.

Claude Code either writes the test or explains that a particular line is unreachable dead code that should be removed, defensive error handling that can't be triggered without a deep system failure, or code requiring a specific infrastructure dependency. You make the call on each one.

The Result

Starting from 34% coverage, you now have over 80%. Total effort: maybe two to three days of guided work with Claude Code — reviewing generated tests, making occasional testability refactors, and running the coverage tool. Without Claude Code, the same effort would take two to three weeks of tedious, soul-crushing test writing.

More importantly, the tests are good. Not just line-coverage padding — they test real behavior, cover meaningful edge cases, and follow consistent patterns. They'll catch real bugs in the future.

Testing Anti-Patterns to Avoid

Even with Claude Code, you can write bad tests. Watch out for these:

Tests that test the implementation, not the behavior. If a test breaks every time you refactor the implementation without changing behavior, it's too tightly coupled. Tell Claude Code: "Test the observable behavior, not the internal implementation."

Tests that are too DRY. Some duplication in tests is fine. Aggressively DRY tests are hard to read and debug. Every test should be understandable on its own without tracing through five layers of shared setup.

Tests without meaningful assertions. A test that calls a function and only checks that it doesn't throw is barely a test. Every test should assert something specific about the output or the side effects.

Flaky tests. Tests that depend on timing, external services, or random data are a maintenance nightmare. Ask Claude Code to make tests deterministic: fixed seeds for random data, mocked clocks for time-dependent code, stubbed external services.

Tests that are too slow. Integration tests are valuable but slow. Keep your unit test suite fast (under thirty seconds for the full suite) and run integration tests separately. Tell Claude Code which kind you want: "Write fast unit tests with mocked dependencies" vs. "Write integration tests using the test database."

Making Testing a Habit

The biggest impact of Claude Code on testing is not any single technique — it's that testing becomes so easy that you actually do it. When writing tests takes five minutes instead of an hour, you stop skipping them. When generating edge case tests is as easy as asking for them, you stop ignoring corner cases.

Build testing into your workflow. After every feature, ask Claude Code to write the tests. After every bug fix, ask it to write a regression test. After every refactor, ask it to verify the existing tests still pass and add any new ones that are needed.

Testing is no longer the tax you pay for writing software. With Claude Code, it's just another step in the development process — and one of the fastest at that.

Chapter 12: Git and Version Control Workflows

Git is the backbone of modern software development. Every feature, every bug fix, every experiment flows through version control. And yet, for many developers, Git remains a source of friction — cryptic commands, merge conflicts that feel like defusing a bomb, commit histories that read like a stream of consciousness.

Claude Code doesn't just write code. It understands Git deeply and handles everything from crafting commit messages to resolving complex merge conflicts. This chapter shows you how to integrate Claude Code into your version control workflow so that Git becomes a smooth, reliable part of your development process instead of an obstacle.

Meaningful Commits with Claude Code

The most basic Git workflow with Claude Code is also one of the most valuable: have it create commits with meaningful, well-structured messages.

The Simple Approach

After making changes (either manually or with Claude Code's help), ask:

> Commit the changes with an appropriate message.

Claude Code will:

1. Run `git status` and `git diff` to understand what has changed
2. Analyze the changes to understand their purpose
3. Write a commit message that describes the *why*, not just the *what*
4. Stage the relevant files and create the commit

The commit messages Claude Code writes are consistently better than what most developers produce under time pressure. Instead of "fix bug" or "update files," you get messages like:

```
Fix race condition in order processing that caused duplicate charges

The processPayment function was not checking whether a payment was already
in progress before initiating a new charge. Added a mutex lock around the
payment processing block and a database check for existing pending payments.
```

Controlling the Commit Style

If your team has a commit message convention, tell Claude Code about it — ideally in your CLAUDE.md file:

```
### Commit Conventions
- Use Conventional Commits format: type(scope): description
- Types: feat, fix, refactor, test, docs, chore, ci
- Keep the first line under 72 characters
- Add a body paragraph for non-trivial changes explaining the "why"
- Reference issue numbers when applicable: Closes #123
```

With this in your CLAUDE.md, every commit Claude Code creates follows the convention:

```
feat(auth): add session expiration and renewal

Sessions now expire after 24 hours of inactivity. When a session is
within 1 hour of expiring, the client automatically requests a renewal.
This prevents users from being unexpectedly logged out during long
work sessions.

Closes #247
```

Selective Staging

Sometimes you've made multiple unrelated changes and want to commit them separately. Claude Code handles this:

> I have changes in both the auth module and the payment module. Create two separate commits: one for the auth changes and one for the payment changes.

Claude Code examines all the modified files, groups them logically, stages each group separately, and creates focused commits for each. This is the kind of Git discipline that improves your project's history but that developers rarely bother with under time pressure.

Amending and Fixing

> I just realized the last commit has a typo in the error message. Fix the typo and amend the commit.

Claude Code finds the typo, fixes it, and amends the previous commit. Just remember the standard rule: never amend a commit that has already been pushed to a shared branch.

Branch Management

Claude Code manages your branches, handling the creation, naming, and switching that forms the backbone of any branching strategy.

Creating Feature Branches

> Create a feature branch for adding email notifications to the order system. Use our branch naming convention.

If your CLAUDE.md specifies a convention like `feature/JIRA-123-short-description`, Claude Code follows it. Otherwise, it creates a sensible name like `feature/order-email-notifications`.

Branch Cleanup

> List all local branches that have been merged into main. Delete them.

Claude Code identifies merged branches and cleans them up, keeping your local repository tidy. It won't delete `main` or the current branch, and it warns you about branches that have unmerged changes.

Pull Request Creation

Creating a pull request is where Claude Code shines. It understands the changes, the context, and the conventions well enough to produce PR descriptions your reviewers will actually appreciate.

Basic PR Creation

> Create a pull request for this branch. Target the main branch.

Claude Code will:

1. Review all commits on the branch since it diverged from main
2. Analyze the changes to understand their scope and purpose
3. Write a PR title and description
4. Create the PR using the GitHub CLI

The resulting PR description includes a summary of the changes, the motivation, any notable implementation decisions, and often a checklist of things the reviewer should pay attention to.

Detailed PR Descriptions

For larger changes, steer the PR description:

> Create a pull request. The description should include: a summary of the changes, the motivation (we needed this for the Q3 launch), a list of the API changes (any new or modified endpoints), testing instructions for the reviewer, and any known limitations.

Claude Code produces a structured, detailed PR description that gives reviewers everything they need. Compare this to the typical PR description: an empty body, or a single sentence like "Added the thing."

PR Templates

If your repository has a PR template (`.github/pull_request_template.md`), Claude Code detects it and fills it in appropriately:

> Create a PR using the repository's PR template. Fill in all sections.

Claude Code reads the template, understands what each section is asking for, and fills them in with relevant information from the changes.

Code Review Assistance

Claude Code is a powerful code review tool, whether you're reviewing someone else's code or preparing your own for review.

Reviewing a PR

> Review the PR at https://github.com/org/repo/pull/142. Focus on correctness, security issues, and potential performance problems.

Claude Code fetches the PR diff, analyzes the changes, and provides a structured review. It catches things like:

- Unvalidated user input that could lead to injection attacks
- Missing error handling for operations that can fail
- Performance issues like N+1 queries or missing database indexes
- Logic errors where the code does not match the apparent intent
- Inconsistencies with the existing codebase patterns
- Missing tests for new functionality

Self-Review Before Submitting

Before creating a PR, ask Claude Code to review your own changes:

> Review all the changes on this branch compared to main. Look for any bugs, security issues, or things I might have missed. Also check for any debugging code or temporary workarounds I might have left in.

A self-review catches embarrassing issues before your teammates see them. Leftover `console.log` statements, TODO comments that should have been resolved, hardcoded test values that should be configuration — Claude Code finds them all.

Responding to Review Comments

When your PR gets review comments, Claude Code can help address them:

> Here are the review comments on my PR: [paste comments or link]. Address each comment by making the requested changes.

Claude Code reads each comment, understands what the reviewer is asking for, and makes the changes. It also drafts response comments explaining what was changed and why.

Resolving Merge Conflicts

Merge conflicts are one of the most stressful parts of Git. You're staring at angle brackets and trying to figure out which version is correct, often in code you didn't write. Claude Code makes this manageable.

Simple Conflict Resolution

> I have merge conflicts after pulling main into my feature branch. Resolve them.

Claude Code reads the conflicted files, understands both sides of each conflict, and resolves them intelligently. It doesn't just pick one side — it merges the intent of both changes when possible. If your branch added a new field to a function and main added a different field, Claude Code keeps both.

Complex Conflicts

For conflicts where the resolution requires understanding the broader context:

> I have merge conflicts in the user service. My branch refactored the error handling, and main added new validation logic. The conflicts are extensive. Resolve them so that both the new error handling pattern and the new validation logic are preserved.

Claude Code reads not just the conflicted hunks but the surrounding code, the commit messages on both sides, and the test files to figure out what each branch was trying to accomplish. The resolution preserves both intentions.

Conflict Prevention

> I'm about to merge main into my feature branch. Before I do, check what files have changed on both branches and warn me about any potential conflicts.

Claude Code compares the two branches and identifies files modified on both sides. The forewarning lets you plan your conflict resolution strategy before you're in the middle of it.

Interactive Rebasing and History Cleanup

A clean Git history makes future debugging and archaeology much easier. Claude Code helps you maintain one.

Squashing Related Commits

> Squash the last 5 commits into a single commit with a good message. These were all part of adding the notification feature.

Claude Code examines the five commits, sees that they're part of a single feature (initial implementation, a bug fix, a style tweak, adding tests, and fixing a test), and squashes them into one clean commit with a message that describes the complete feature.

Reordering and Cleaning

> Clean up the commit history on this branch. Group related changes together, squash fixup commits into their parent commits, and reword any unclear commit messages.

Claude Code analyzes the commit history, groups commits that logically belong together, and produces a clean, readable history.

Splitting Commits

Sometimes a commit does too much:

> The commit "update user module" actually contains three unrelated changes: a bug fix, a new feature, and a refactor. Split it into three separate commits.

Claude Code examines the changes, groups them by purpose, and creates three focused commits.

Git Hooks with Claude Code

Git hooks automate checks and tasks at various points in the Git workflow. Claude Code can help you set them up and work with them.

Pre-Commit Hooks

> Set up a pre-commit hook that runs the linter and type checker on staged files. If either fails, block the commit.

Claude Code creates the hook script, makes it executable, and tests it. If you're using a framework like Husky or pre-commit, Claude Code works within that framework.

Commit Message Hooks

> Set up a commit-msg hook that validates commit messages against our Conventional Commits format. Reject commits that do not match the pattern.

Claude Code writes a hook that parses the commit message and validates it against the convention, returning a helpful error message when the format is wrong.

Working With Existing Hooks

When a commit fails because of a pre-commit hook, Claude Code can fix the issue:

> The pre-commit hook failed because of linting errors. Fix the linting errors and commit again.

Claude Code reads the hook output, identifies the linting violations, fixes them, and retries the commit.

Real-World Workflow: Managing a Feature Branch

Let's walk through a complete feature branch lifecycle, from creation to merge, using Claude Code at every step.

Day 1: Branch Creation and Initial Work

Start by creating the branch:

> Create a new branch for implementing the user notification preferences feature. We're working on JIRA ticket NOTIFY-234.

Claude Code creates `feature/NOTIFY-234-notification-preferences` and switches to it.

Build the initial implementation:

> Implement the notification preferences feature. Users should be able to choose which types of notifications they receive (email, SMS, push) and set quiet hours during which no notifications are sent. Start with the backend: database schema, API endpoints, and service logic.

Claude Code generates the migration, models, services, routes, and tests. You review and iterate.

Commit the work:

> Commit the backend implementation with an appropriate message.

```
feat(notifications): add notification preferences backend

Add user notification preferences with support for email, SMS, and push
channels. Users can enable/disable each channel and set quiet hours
(start time, end time, timezone) during which notifications are held.

- Migration: add notification_preferences table
- API: CRUD endpoints at /api/users/:id/notification-preferences
- Service: preference management with quiet hours logic
- Tests: unit and integration tests for all endpoints

Refs: NOTIFY-234
```

Day 2: Frontend and Refinement

Continue the work:

> Now add the frontend: a notification preferences page accessible from the user settings. Include toggles for each channel and a time picker for quiet hours.

Review, iterate, and commit:

> Commit the frontend changes.

Day 3: Pulling in Updates and Handling Conflicts

Main has moved on. Time to sync:

> Pull the latest changes from main into our branch. Resolve any conflicts.

Claude Code fetches main, merges it, and resolves any conflicts. If the conflicts hit areas relevant to your feature, it makes sure the resolution preserves your changes while incorporating the updates from main.

Day 4: Self-Review and PR Preparation

Before creating the PR, do a self-review:

> Review all changes on this branch compared to main. Check for any issues, missing tests, or cleanup needed before I submit a PR.

Claude Code reviews the full diff, identifies a few issues — a missing error handler on one endpoint, a test that doesn't clean up its database state, and a TODO comment that should be resolved — and fixes them.

Clean up the history:

> Squash the fixup commits and clean up the commit history so it tells a clear story.

Claude Code produces a clean history: one commit for the backend, one for the frontend, and one for the final polish.

Day 5: PR Creation and Review Response

Create the PR:

> Create a pull request targeting main. Include a thorough description with the motivation, summary of changes, API changes, and testing instructions.

Claude Code creates the PR with a detailed description. Reviewers see exactly what changed, why, and how to test it.

Handle review feedback:

> The reviewer left these comments: [paste]. Address each one, then push the changes.

Claude Code makes the requested changes, commits them with appropriate messages, and pushes.

Day 6: Final Merge

The PR is approved. Time to merge:

> The PR is approved. Rebase on main to make sure we're up to date, then merge using a squash merge.

Claude Code rebases, confirms there are no conflicts, and completes the merge. The feature is in main with a clean history.

The Complete Picture

This workflow used Claude Code for:

- Branch creation with proper naming
- Feature implementation across many files
- Meaningful commits following team conventions
- Merging updates from main with conflict resolution
- Self-review before PR submission
- Commit history cleanup
- PR creation with detailed descriptions
- Addressing review feedback
- Final rebase and merge

Every one of these steps is something you can do manually. But having Claude Code handle the mechanical parts — staging, message writing, conflict resolution, PR descriptions — frees you to focus on the decisions that matter: architecture, logic, code quality.

Git Best Practices with Claude Code

Commit often, commit meaningfully. Because Claude Code makes committing easy, there's no excuse for massive, monolithic commits. Commit after each logical unit of work.

Keep your CLAUDE.md Git conventions up to date. If your team switches from merge commits to squash merges, update the CLAUDE.md. Claude Code follows whatever convention you specify.

Review before you push. Always ask Claude Code to review changes before pushing. A quick self-review catches issues that would embarrass you in a PR.

Use branches for everything. Claude Code makes branch management trivial. There's no reason to commit experimental work directly to main.

Write PR descriptions. With Claude Code generating them, there's no excuse for empty PR descriptions. Good descriptions make reviews faster and your Git history more useful.

Git is a power tool, and like any power tool, it's most effective when used skillfully. Claude Code doesn't replace your understanding of Git — it handles the tedious parts so you can focus on the strategic decisions that make your version control workflow smooth and your project history clean.

Part IV: Extending Claude Code

"The best tool is the one you build yourself — with a better tool."

— Anonymous

Chapter 13: Hooks and Automation

You've been using Claude Code as a conversational tool — you ask it to do things, it does them, and you review the results. That workflow is powerful, but it's also manual. Every time Claude Code writes a file, you might want it auto-formatted. Every time it finishes a long task, you might want a Slack ping. Every time it tries to run a destructive command, you might want a safety check.

Hooks give you all of that. They're the mechanism for injecting your own automation into Claude Code's workflow, turning it from a tool you supervise into a tool that operates within guardrails you define. If CLAUDE.md is how you teach Claude Code what to do, hooks are how you enforce what happens around what it does.

This chapter covers what hooks are, how they work, and how to build custom workflows that make Claude Code fit seamlessly into your development process.

What Are Hooks?

Hooks are shell commands that Claude Code executes automatically when specific events occur. They're configured in your `settings.json` file — either at the project level (`.claude/settings.json`) or the user level (`~/.claude/settings.json`) — and they run outside of Claude's decision-making. Claude doesn't choose whether to run a hook. If you've configured a hook for an event, it fires every time that event occurs.

This distinction matters. Hooks are not prompts. They're not suggestions. They're deterministic automation: when X happens, run Y. That makes them reliable in a way natural-language instructions cannot be. You'd never want "please remember to run Prettier after every file edit" in your CLAUDE.md when you can guarantee it with a hook.

The basic structure of a hook configuration looks like this:

```
{
  "hooks": {
    "PreToolUse": [
```

```
      {
        "matcher": "Write",
        "command": "echo 'A file is about to be written'"
      }
    ]
  }
}
```

Each hook has an event type, an optional matcher that narrows which tool invocations trigger it, and a command that runs in your system shell. The command receives context about the event via environment variables and standard input, and its exit code determines what happens next.

Hook Events

Claude Code exposes several event types you can attach hooks to. Understanding when each fires is essential to building effective automation.

PreToolUse

This fires before Claude Code executes a tool call. The tool has been planned but not yet run. Your hook receives the tool name and its input parameters via stdin as a JSON object. If your hook exits with code `2`, the tool call is blocked — Claude Code won't execute it and will see your stderr as the error message. (Other non-zero codes log a hook error but don't block.)

PreToolUse is your safety net. It's where you enforce invariants: "never delete files in the `src/` directory," "never run `rm -rf` with a path shorter than three segments," "always reject writes to `.env` files." Because it fires before execution, you prevent mistakes rather than clean up after them.

```
{
  "hooks": {
    "PreToolUse": [
      {
        "matcher": "Bash",
        "command": "python3 .claude/hooks/validate-command.py"
      }
    ]
```

```
  }
}
```

The matcher here is "Bash", meaning this hook fires only when Claude Code is about to execute a shell command. The Python script receives the command details on stdin, inspects them, and exits with code 2 if the command should be blocked. Any output your hook writes to stderr becomes the error message Claude Code sees.

PostToolUse

This fires after a tool call completes. Your hook receives the tool name, its input, and its output. The exit code of a PostToolUse hook doesn't block anything — the tool has already run — but you can use it for side effects: formatting files, logging actions, updating dashboards, or triggering downstream processes.

PostToolUse is where most practical automation lives. The canonical example is auto-formatting:

```
{
  "hooks": {
    "PostToolUse": [
      {
        "matcher": "Write|Edit",
        "command": "python3 .claude/hooks/auto-format.py"
      }
    ]
  }
}
```

The matcher "Write|Edit" fires this hook whenever Claude Code writes or edits a file. The formatting script reads the file path from the hook's input JSON, runs your formatter (Prettier, Black, gofmt, whatever your project uses), and the formatted result ends up on disk.

Notification

This fires when Claude Code produces a notification — typically when a long-running task completes or when Claude Code needs your attention. Notification hooks are ideal for plugging Claude Code into your communication tools.

```
{
  "hooks": {
```

```
    "Notification": [
      {
        "command": "python3 .claude/hooks/slack-notify.py"
      }
    ]
  }
}
```

No matcher is needed here because Notification events don't have tool-specific subtypes. Every notification triggers the hook.

Stop

This fires when Claude Code is about to stop its turn — when it believes it has completed the task. A Stop hook can inspect the final state and, if it returns exit code `2` with a message on stderr, force Claude Code to continue working. Powerful for enforcing quality gates: "don't stop until all tests pass," or "don't stop until the linter is clean." A plain `exit 1` won't do it — only `exit 2` is blocking.

```
{
  "hooks": {
    "Stop": [
      {
        "command": "python3 .claude/hooks/check-tests.py"
      }
    ]
  }
}
```

Anatomy of a Hook Command

Your hook command runs in a standard shell environment. Claude Code provides context through environment variables and stdin.

Environment variables available to your hook include:

- `CLAUDE_PROJECT_DIR` — the root directory of the project. Set for all hooks. This is the one you'll actually use.

- `CLAUDE_PLUGIN_ROOT` — the root directory of the plugin. Set only for hooks running inside a plugin.
- `CLAUDE_PLUGIN_DATA` — the plugin's data directory. Plugin hooks only.
- `CLAUDE_CODE_REMOTE` — set to `"true"` when Claude Code is running in a remote web environment; not set locally.
- `CLAUDE_ENV_FILE` — path to an env file your hook can write to. Available in `SessionStart`, `Setup`, `CwdChanged`, and `FileChanged` hooks.

Notably absent: there is no `CLAUDE_TOOL_NAME`, no `CLAUDE_EVENT`, no `CLAUDE_SESSION_ID`. Those values exist, but they come in on stdin, not as env vars.

Stdin receives a JSON object with the full details of the event. Every hook gets a common set of fields, plus event-specific extras. The common fields:

```
{
  "session_id": "abc123",
  "transcript_path": "/path/to/transcript.jsonl",
  "cwd": "/current/working/directory",
  "permission_mode": "default",
  "hook_event_name": "PreToolUse"
}
```

For `PreToolUse` and `PostToolUse`, the payload also includes the tool name and the tool's input (and, for `PostToolUse`, its output). For subagent hooks, you also get `agent_id` and `agent_type`. The exact schema varies by event type, but you can always parse it as JSON.

Read it from stdin in your hook script:

```
#!/usr/bin/env python3
import json, sys

payload = json.load(sys.stdin)
tool_name = payload["tool_name"]
session_id = payload["session_id"]
```

If you prefer bash, pipe through `jq`:

```
INPUT=$(cat)
TOOL_NAME=$(echo "$INPUT" | jq -r .tool_name)
```

HTTP hooks receive the same JSON payload as a POST body and signal their result via HTTP status code instead of an exit code.

Exit codes matter, and the rule is narrower than people assume: only **exit 2** is blocking. Anything else non-zero is a non-blocking error — the transcript shows a hook-error notice with the first line of stderr, and execution continues.

- Exit code 0: success, proceed normally
- Exit code 2 (PreToolUse): block the tool call; stderr is fed back to Claude as the reason
- Exit code 2 (Stop): prevent Claude Code from stopping; stderr becomes the continuation prompt
- Exit code 2 (UserPromptSubmit / PreCompact / other blocking events): same pattern — block the action, stderr explains why
- Other non-zero exit codes (including the reflexive `exit 1`): non-blocking error. Logged, not enforced.

Per the docs, verbatim: "Exit code 1 is treated as a non-blocking error, not a blocking error. If your hook is meant to enforce a policy, use `exit 2`."

Stdout output from your hook is passed back to Claude Code as additional context. Useful for PreToolUse hooks that want to modify or annotate the operation, or PostToolUse hooks that want to report results.

Building a Pre-Tool-Call Safety Gate

Let's build a practical PreToolUse hook that prevents Claude Code from running dangerous shell commands. The kind of safety net every team should have.

Create a file at `.claude/hooks/validate-bash.py`:

```
#!/usr/bin/env python3
import sys
import json

BLOCKED_PATTERNS = [
    "rm -rf /",
    "rm -rf ~",
    "DROP DATABASE",
```

```
    "DROP TABLE",
    "> /dev/sda",
    "mkfs.",
    ":(){:|:&};:",
    "chmod -R 777 /",
]

def main():
    event = json.load(sys.stdin)
    command = event.get("tool_input", {}).get("command", "")

    for pattern in BLOCKED_PATTERNS:
        if pattern.lower() in command.lower():
            print(f"BLOCKED: Command contains dangerous pattern '{pattern}'",
                  file=sys.stderr)
            sys.exit(2)  # exit 2 blocks; exit 1 would only log a warning

    sys.exit(0)

if __name__ == "__main__":
    main()
```

Then configure it in `.claude/settings.json`:

```
{
  "hooks": {
    "PreToolUse": [
      {
        "matcher": "Bash",
        "command": "python3 .claude/hooks/validate-bash.py"
      }
    ]
  }
}
```

Now, if Claude Code ever tries to run a command containing `rm -rf /` or `DROP DATABASE`, the hook blocks it before execution. Claude Code sees the error message and adjusts its approach. Defense in depth — you're not relying on Claude Code's judgment alone.

Building a Post-Tool-Call Formatter

Auto-formatting is the most common PostToolUse hook, and for good reason. Claude Code generates code that's functionally correct but doesn't always match your formatting preferences exactly. A PostToolUse hook that runs your formatter on every written file eliminates the friction entirely.

A hook script that handles multiple languages:

```
#!/usr/bin/env python3
import sys
import json
import subprocess
import os

FORMATTERS = {
    ".js": ["npx", "prettier", "--write"],
    ".ts": ["npx", "prettier", "--write"],
    ".jsx": ["npx", "prettier", "--write"],
    ".tsx": ["npx", "prettier", "--write"],
    ".py": ["python3", "-m", "black"],
    ".go": ["gofmt", "-w"],
    ".rs": ["rustfmt"],
}

def main():
    event = json.load(sys.stdin)
    file_path = event.get("tool_input", {}).get("file_path", "")

    if not file_path or not os.path.exists(file_path):
        sys.exit(0)

    _, ext = os.path.splitext(file_path)
    formatter = FORMATTERS.get(ext)

    if formatter:
        try:
            subprocess.run(
                formatter + [file_path],
                capture_output=True,
                timeout=10,
            )
        except (subprocess.TimeoutExpired, FileNotFoundError):
            pass  # Formatter not available; skip silently
```

```
    sys.exit(0)

if __name__ == "__main__":
    main()
```

Configure it:

```
{
  "hooks": {
    "PostToolUse": [
      {
        "matcher": "Write|Edit",
        "command": "python3 .claude/hooks/auto-format.py"
      }
    ]
  }
}
```

Every file Claude Code writes or edits now gets formatted automatically. No more "hey, can you run Prettier on that?" follow-ups.

Building a Notification Hook

When Claude Code is working on a long task — say, refactoring a hundred files or running a full test suite — you probably don't want to stare at your terminal waiting. A notification hook lets you walk away and get pinged when Claude needs your attention.

A simple hook that sends a Slack message via a webhook:

```
#!/usr/bin/env python3
import sys
import json
import urllib.request

SLACK_WEBHOOK_URL = "https://hooks.slack.com/services/YOUR/WEBHOOK/URL"

def main():
    event = json.load(sys.stdin)
    message = event.get("message", "Claude Code needs your attention")
```

```python
    payload = json.dumps({
        "text": f":robot_face: Claude Code: {message}"
    })

    req = urllib.request.Request(
        SLACK_WEBHOOK_URL,
        data=payload.encode("utf-8"),
        headers={"Content-Type": "application/json"},
    )

    try:
        urllib.request.urlopen(req, timeout=5)
    except Exception:
        pass  # Don't block on notification failures

    sys.exit(0)

if __name__ == "__main__":
    main()
```

You could also send a desktop notification using `osascript` on macOS, `notify-send` on Linux, or `powershell` on Windows. The mechanism doesn't matter — what matters is that you're no longer tethered to your terminal.

Building a Quality Gate with Stop Hooks

Stop hooks are the most powerful hook type because they can force Claude Code to keep working. They let you create automated quality gates that ensure Claude Code doesn't declare "done" until your standards are met.

A Stop hook that checks whether tests pass:

```python
#!/usr/bin/env python3
import sys
import subprocess
import os

PROJECT_DIR = os.environ.get("CLAUDE_PROJECT_DIR", ".")

def main():
    result = subprocess.run(
        ["npm", "test"],
```

```python
        capture_output=True,
        text=True,
        cwd=PROJECT_DIR,
        timeout=120,
    )

    if result.returncode != 0:
        print(
            f"Tests are failing. Please fix them before stopping.\n\n"
            f"Test output:\n{result.stdout[-2000:]}\n{result.stderr[-2000:]}",
            file=sys.stderr,
        )
        sys.exit(2)  # exit 2 forces Claude Code to keep working

    sys.exit(0)

if __name__ == "__main__":
    main()
```

With this hook, when Claude Code says "I'm done," the hook runs `npm test`. If tests fail, Claude Code receives the test output and keeps working to fix them. The result is a tight loop: write code, try to stop, tests fail, fix the code, try to stop again, tests pass, actually stop.

Be careful with Stop hooks. A poorly written one can create infinite loops where Claude Code never finishes. Always include escape hatches — only check tests if Claude Code actually modified source files, or limit the number of retry cycles.

Combining Hooks into Workflows

Individual hooks are useful. Combined hooks are transformative. A complete `settings.json` that implements a professional development workflow:

```json
{
  "hooks": {
    "PreToolUse": [
      {
        "matcher": "Bash",
        "command": "python3 .claude/hooks/validate-bash.py"
      },
      {
```

```
        "matcher": "Write|Edit",
        "command": "python3 .claude/hooks/check-write-permissions.py"
      }
    ],
    "PostToolUse": [
      {
        "matcher": "Write|Edit",
        "command": "python3 .claude/hooks/auto-format.py"
      },
      {
        "matcher": "Write|Edit",
        "command": "python3 .claude/hooks/auto-lint.py"
      },
      {
        "matcher": "Bash",
        "command": "python3 .claude/hooks/log-commands.py"
      }
    ],
    "Notification": [
      {
        "command": "python3 .claude/hooks/slack-notify.py"
      }
    ],
    "Stop": [
      {
        "command": "python3 .claude/hooks/check-tests.py"
      }
    ]
  }
}
```

This configuration enforces a workflow where:

1. Shell commands are validated before execution (no dangerous patterns)
2. File writes are checked against a permission list (no writing to protected paths)
3. Every written file is auto-formatted and auto-linted
4. Every shell command is logged for audit purposes
5. You get a Slack ping whenever Claude Code needs attention
6. Claude Code cannot stop until tests pass

That's a professional-grade development workflow, and it requires zero manual intervention. You set it up once, and it runs forever.

Tips for Writing Effective Hooks

Keep hooks fast. Every hook adds latency to Claude Code's workflow. A PreToolUse hook that takes five seconds to run means Claude Code waits five seconds before every tool call. Aim for sub-second execution. If you need to do something slow, consider whether it needs to block execution or whether a PostToolUse hook (which doesn't block) would suffice.

Handle errors gracefully. Your hook should never crash in a way that produces confusing error messages. Catch exceptions, write clear error messages on stderr, and use meaningful exit codes.

Use the project's `.claude/hooks/` directory. This keeps hook scripts versioned with the project and shared across the team. Everyone who clones the repo gets the same automation.

Test hooks independently. You can test a hook by piping JSON to its stdin and checking its exit code. You don't need Claude Code running to verify your hooks work:

```
echo '{"tool_input": {"command": "rm -rf /"}}' | python3 .claude/hooks/validate-bash.py
echo $?  # Should be 2 (blocked)
```

Be careful with PreToolUse blockers. If your PreToolUse hook is too aggressive, Claude Code gets stuck in a loop where it keeps trying things that get blocked. Make your error messages specific so Claude Code can adjust its approach.

Version your hook configurations. The `settings.json` file in `.claude/` should be committed to version control. That ensures the entire team uses the same hooks and that your automation is reproducible.

What Hooks Cannot Do

Hooks are powerful, but they have boundaries. They cannot modify Claude Code's prompts or influence its reasoning. They cannot inject tool results or fake tool outputs. They cannot change which model Claude Code uses or alter its system prompt. They operate at the boundary between Claude Code's decisions and their execution.

Think of hooks as middleware in a web framework. They intercept requests and responses, but they don't change the application logic. They're the right tool for automation, validation, and side effects — not for changing how Claude Code thinks.

Summary

Hooks transform Claude Code from a conversational tool into an automated development system. PreToolUse hooks enforce safety invariants. PostToolUse hooks handle formatting, linting, and logging. Notification hooks keep you informed. Stop hooks enforce quality gates.

The key insight is that hooks are deterministic. They don't depend on Claude Code remembering your instructions or interpreting them correctly. They run every time, exactly as configured. For anything that must always happen — formatting, validation, safety checks — hooks are the right mechanism.

In the next chapter, we'll extend Claude Code's capabilities further by connecting it to external tools and services through MCP servers.

Chapter 14: MCP Servers and External Tools

Claude Code is capable within its default boundaries — it can read files, write code, run shell commands, and search your codebase. But real-world development involves more than files on disk. You query databases. You check deployment status. You look up customer data. You consult monitoring dashboards. You hit third-party APIs.

MCP — the Model Context Protocol — is how you extend Claude Code's reach beyond the filesystem. It's an open standard developed by Anthropic that lets AI tools connect to external services through a consistent interface. Configure an MCP server and you give Claude Code a new set of tools it can use as naturally as it reads files or runs shell commands.

This chapter covers what MCP is, how it works with Claude Code, which servers are worth using, and how to build your own.

What Is MCP?

The Model Context Protocol is a standard way for AI applications to talk to external tools and data sources. Think of it as a USB port for AI — a universal interface that lets any AI application talk to any service, as long as both sides speak the protocol.

Before MCP, every AI tool had to build custom integrations with every service. Want your AI to talk to PostgreSQL? Write a PostgreSQL integration. Want it to talk to Jira? Write a Jira integration. This doesn't scale — there are thousands of services and dozens of AI tools, and building N-times-M integrations is impractical.

MCP solves this with a standard protocol. A service implements an MCP server once, and every MCP-compatible AI tool can use it. An AI tool implements MCP client support once, and it can talk to every MCP server. Claude Code is an MCP client. The servers you connect provide the tools.

The protocol is straightforward. An MCP server declares what tools it offers (their names, parameters, and descriptions), and the AI client can invoke those tools during conversation. The server runs locally on your machine or on a remote host, communicates over stdio or HTTP, and returns structured results the AI can reason about.

How MCP Works in Claude Code

When you configure an MCP server in Claude Code, the server starts as a subprocess (for stdio-based servers) or connects via HTTP (for remote servers). Claude Code queries the server for its available tools at startup, and those tools appear alongside Claude Code's built-in capabilities. When Claude Code decides it needs one of those tools — say, querying a database — it sends a tool call to the MCP server, which executes the operation and returns the result.

From your perspective, this is seamless. You tell Claude Code "check how many users signed up yesterday" and it uses the database MCP server to run the query, as naturally as it would run `grep` to search your code.

MCP servers are configured in your settings file. You can set them up at the project level (`.claude/settings.json`), the user level (`~/.claude/settings.json`), or both. Project-level servers fit things specific to a project — its database, its deployment system. User-level servers fit personal tools — your email, your calendar, your note system.

Here is the basic configuration structure:

```
{
  "mcpServers": {
    "server-name": {
      "command": "npx",
      "args": ["-y", "@modelcontextprotocol/server-postgres"],
      "env": {
        "POSTGRES_CONNECTION_STRING": "postgresql://user:pass@localhost:5432/myd
b"
      }
    }
  }
}
```

Each server gets a name (which you choose), a command to start it, optional arguments, and optional environment variables. When Claude Code launches, it starts all configured MCP servers and discovers their tools.

Built-in vs. Custom MCP Servers

The MCP ecosystem has two layers. The first is official and community-maintained servers for popular services — databases, APIs, development tools. Ready to install and use. The second layer is custom servers you build yourself for your specific needs.

Official and Popular MCP Servers

The MCP ecosystem has grown rapidly. Here are the servers most relevant to development workflows:

PostgreSQL (`@modelcontextprotocol/server-postgres`): Lets Claude Code query your PostgreSQL databases directly. It can run SELECT queries, describe schemas, list tables, and inspect data. This changes backend development — instead of switching to a database client to check your schema, you just ask Claude Code.

```
{
  "mcpServers": {
    "postgres": {
      "command": "npx",
      "args": ["-y", "@modelcontextprotocol/server-postgres"],
      "env": {
        "POSTGRES_CONNECTION_STRING": "postgresql://dev:devpass@localhost:5432/m
yapp_dev"
      }
    }
  }
}
```

GitHub (`@modelcontextprotocol/server-github`): Gives Claude Code access to GitHub's API — issues, pull requests, repositories, actions. You can ask Claude Code to check the status of a PR, list open issues, or review CI results without leaving your terminal.

Filesystem (`@modelcontextprotocol/server-filesystem`): Extends Claude Code's file access beyond the current project. Useful when you need Claude Code to reference files in other directories — documentation repositories, shared configuration, design assets.

SQLite (`@modelcontextprotocol/server-sqlite`): Like the PostgreSQL server but for SQLite databases. Good for mobile development or applications that use SQLite for local storage.

Brave Search (`@modelcontextprotocol/server-brave-search`): Gives Claude Code web search. When it encounters an unfamiliar library or needs to look up current documentation, it can search the web directly.

Memory (`@modelcontextprotocol/server-memory`): Provides persistent key-value storage that survives across sessions. Handy for project-specific context that Claude Code should remember between conversations.

Slack (`@modelcontextprotocol/server-slack`): Lets Claude Code interact with Slack — reading messages, posting updates, searching conversation history. Useful when development context lives in Slack threads.

Installing an MCP Server

Most MCP servers are npm packages. You don't need to install them globally — `npx -y` in the configuration command downloads and runs them on the fly. For Python-based servers, use `uvx` or `python -m` instead.

The typical setup flow is:

1. Find the server you want (check the MCP server directory at modelcontextprotocol.io)
2. Add its configuration to your settings.json
3. Provide any necessary credentials via the `env` field
4. Restart Claude Code
5. Verify the server loaded by asking Claude Code what tools it has

If a server fails to start, Claude Code will show a warning at startup. Common issues include missing credentials, network connectivity problems, or version mismatches.

Connecting to Databases

Database access is the highest-impact MCP integration for most developers. Asking Claude Code "what does the users table look like?" or "how many orders were placed in the last hour?" while you write code is a workflow improvement that, once you've felt it, is hard to give up.

Setting Up Database Access

For PostgreSQL:

```
{
  "mcpServers": {
    "staging-db": {
      "command": "npx",
      "args": ["-y", "@modelcontextprotocol/server-postgres"],
      "env": {
        "POSTGRES_CONNECTION_STRING": "postgresql://readonly:password@staging.ex
ample.com:5432/app"
      }
    }
  }
}
```

For MySQL, there are community MCP servers available, or you can use a generic SQL server. For MongoDB, there is an official MCP server as well.

Best Practices for Database MCP Servers

Always use read-only credentials. You don't want Claude Code running `DELETE FROM users` on your production database. Create a database user with SELECT-only permissions and use that for your MCP connection.

Connect to staging, not production. Even with read-only access, querying production databases has performance implications. Use a staging or development replica.

Use connection pooling thoughtfully. The MCP server holds a connection to your database for the duration of the Claude Code session. If you run many sessions, this adds up.

Be mindful of sensitive data. Query results flow through Claude Code's conversation, which means they are sent to Anthropic's API. Don't query tables containing passwords, API keys, or personally identifiable information unless you're comfortable with that data flowing through the API.

Connecting to APIs and Services

Beyond databases, MCP servers can connect Claude Code to almost any API or service. The pattern is the same: configure a server, and Claude Code gets new tools.

Example: GitHub Integration

```
{
  "mcpServers": {
    "github": {
      "command": "npx",
      "args": ["-y", "@modelcontextprotocol/server-github"],
      "env": {
        "GITHUB_PERSONAL_ACCESS_TOKEN": "ghp_your_token_here"
      }
    }
  }
}
```

With this configured, you can ask Claude Code things like:

- "What are the open PRs on this repo?"
- "Show me the comments on PR #142."
- "What CI checks are failing on the main branch?"
- "Create an issue for the bug we just found."

Example: Web Search

```
{
  "mcpServers": {
    "search": {
      "command": "npx",
      "args": ["-y", "@modelcontextprotocol/server-brave-search"],
      "env": {
```

```
        "BRAVE_API_KEY": "your_brave_api_key"
      }
    }
  }
}
```

Web search is especially useful when Claude Code hits an unfamiliar error message, needs to look up the current version of a library, or wants to check documentation for a tool it hasn't seen in its training data.

Building a Custom MCP Server

Sometimes the server you need doesn't exist. Maybe your company has an internal deployment system, a custom monitoring stack, or a proprietary API that Claude Code needs to access. In those cases, you build your own.

MCP servers can be built in any language that can read and write JSON over stdio. The most common choices are TypeScript (using the `@modelcontextprotocol/sdk` package) and Python (using the `mcp` package).

A Simple MCP Server in TypeScript

Here's a minimal MCP server that provides a tool to check the status of a deployment:

```
import { McpServer } from "@modelcontextprotocol/sdk/server/mcp.js";
import { StdioServerTransport } from "@modelcontextprotocol/sdk/server/stdio.js";
import { z } from "zod";

const server = new McpServer({
  name: "deploy-status",
  version: "1.0.0",
});

server.tool(
  "check_deployment",
  "Check the current deployment status for an environment",
  {
    environment: z.enum(["staging", "production"]).describe(
      "The environment to check"
    ),
```

```
  },
  async ({ environment }) => {
    // In reality, this would call your deployment API
    const response = await fetch(
      `https://deploy.internal.example.com/api/status/${environment}`,
      {
        headers: {
          Authorization: `Bearer ${process.env.DEPLOY_API_TOKEN}`,
        },
      }
    );

    const status = await response.json();

    return {
      content: [
        {
          type: "text",
          text: JSON.stringify(status, null, 2),
        },
      ],
    };
  }
);

const transport = new StdioServerTransport();
await server.connect(transport);
```

This server declares a single tool (`check_deployment`) that takes an environment parameter and returns the deployment status. You register it with Claude Code like any other MCP server:

```
{
  "mcpServers": {
    "deploy": {
      "command": "node",
      "args": ["./tools/deploy-status-server.mjs"],
      "env": {
        "DEPLOY_API_TOKEN": "your_token_here"
      }
    }
  }
}
```

Now you can ask Claude Code "what's the status of the staging deployment?" and it will use your custom tool to find out.

A Simple MCP Server in Python

The Python SDK is equally straightforward:

```python
from mcp.server.fastmcp import FastMCP

mcp = FastMCP("deploy-status")

@mcp.tool()
async def check_deployment(environment: str) -> str:
    """Check the current deployment status for an environment.

    Args:
        environment: The environment to check (staging or production)
    """
    import httpx
    import os

    async with httpx.AsyncClient() as client:
        response = await client.get(
            f"https://deploy.internal.example.com/api/status/{environment}",
            headers={
                "Authorization": f"Bearer {os.environ['DEPLOY_API_TOKEN']}"
            },
        )
        return response.text

if __name__ == "__main__":
    mcp.run(transport="stdio")
```

Configure it:

```json
{
  "mcpServers": {
    "deploy": {
      "command": "python3",
      "args": ["./tools/deploy_status_server.py"],
      "env": {
        "DEPLOY_API_TOKEN": "your_token_here"
      }
    }
```

```
  }
}
```

Design Tips for Custom MCP Servers

Give tools clear descriptions. Claude Code uses the description to decide when to invoke a tool. A tool described as "check deployment status" gets called when you ask about deployments. A tool described as "do thing" won't be used correctly.

Use specific parameter schemas. The more you constrain parameters with types, enums, and descriptions, the more reliably Claude Code will call your tools correctly.

Return structured data. JSON results are easier for Claude Code to reason about than free-form text. Include the relevant data and nothing extra.

Handle errors gracefully. If your tool fails, return an error message Claude Code can understand and relay to you, rather than crashing the server.

Keep servers focused. One server per domain is a good rule. A deployment server, a monitoring server, a customer-data server — not one mega-server that does everything.

Security Considerations

MCP servers run on your machine with your credentials. That makes them powerful but also introduces security surface area. The key considerations:

Credential management. MCP server configurations often include API tokens and database passwords. Never commit these to version control. Use environment variables, a secrets manager, or a `.env` file in your `.gitignore`.

A safer pattern for project settings is to reference environment variables that each developer sets locally:

```
{
  "mcpServers": {
    "postgres": {
      "command": "npx",
```

```
      "args": ["-y", "@modelcontextprotocol/server-postgres"],
      "env": {
        "POSTGRES_CONNECTION_STRING": "${DB_CONNECTION_STRING}"
      }
    }
  }
}
```

Principle of least privilege. Give MCP servers the minimum permissions they need. Database servers get read-only access. GitHub servers get a token scoped to the necessary repositories. Deployment servers get read-only status access, not deploy permissions.

Network exposure. Stdio-based MCP servers communicate only through local pipes and don't expose network ports. HTTP-based servers do. If you use HTTP-based servers, bind them to localhost only and consider authentication.

Data flow awareness. Everything an MCP server returns to Claude Code becomes part of the conversation, which is processed by Anthropic's API. Watch what data flows through MCP tools, especially with database servers that might return customer data.

Audit MCP server code. If you use community MCP servers, read their source. They run on your machine with whatever credentials you provide. Treat them like any other dependency — review before trusting.

Real-World Workflow: Claude Code Connected to Your Staging Database

Let's walk through a realistic workflow that shows the power of MCP integration. You're building a feature that adds user activity tracking to your application. You've configured the PostgreSQL MCP server to connect to your staging database.

Step 1: Understanding the existing schema.

You start by asking Claude Code to explore the relevant tables:

> "Show me the schema for the users table and any related tables."

Claude Code uses the PostgreSQL MCP server to run schema queries. It returns the table structure, column types, indexes, and foreign key relationships. No switching to a database client, no remembering the exact `\d+` syntax.

Step 2: Designing the new table.

> "I need to add an activity_events table that tracks user actions. Each event should have a user_id, event_type, metadata JSON field, and a timestamp. Design the migration."

Claude Code already knows your schema from step 1. It generates a migration file that creates the table with appropriate types, a foreign key to users, and indexes on user_id and timestamp. Column types match what it saw in your existing schema — if you use `timestamptz` elsewhere, it uses that here too.

Step 3: Validating against real data.

After you run the migration on staging, you ask Claude Code to verify:

> "Check that the activity_events table was created correctly and show me a sample of the data after I seed it."

Claude Code queries the staging database to confirm the schema matches the migration and shows sample rows.

Step 4: Writing queries for the application.

> "Write the data access layer for activity events. I need functions to insert events, query events by user, and get event counts by type for the last 30 days."

Claude Code writes the data access code, and because it has seen your actual database schema via MCP, the column names, types, and relationships are right. No guessing, no placeholder names.

Step 5: Verifying query performance.

> "Run an EXPLAIN on the 'events by user in the last 30 days' query with a real user_id from staging."

Claude Code runs EXPLAIN ANALYZE through the MCP server, shows you the query plan, and tells you whether your indexes are being used. If a query is doing a sequential scan, it suggests adding an index.

This workflow is tighter than the traditional cycle of write-code, switch-to-db-client, check-schema, switch-back, write-more-code. Everything stays in one conversation, and Claude Code has full context at every step.

Managing Multiple MCP Servers

As you add more MCP servers, organization matters. Some patterns:

Separate project and personal servers. Your project's `.claude/settings.json` should contain servers the whole team uses (the project database, the project's GitHub integration). Your personal `~/.claude/settings.json` should contain servers that are just for you (your personal note system, your calendar).

Use descriptive server names. Name servers by what they connect to, not what they do: `staging-db` rather than `database`, `github-myorg` rather than `github`. This matters most when you have multiple instances of the same server type.

Document server setup. If your project uses MCP servers, document the setup in your CLAUDE.md or a project README. List what servers are needed, what credentials to obtain, and how to configure them.

Summary

MCP servers extend Claude Code from a tool that works with your code to a tool that works with your entire development infrastructure. Databases, APIs, deployment systems, monitoring tools — anything with an API can be connected to Claude Code through MCP.

The ecosystem of pre-built servers covers the most common needs, and building a custom server for your specific infrastructure is straightforward. The key is being thoughtful about security — read-only credentials, limited permissions, and awareness of what data flows through the system.

In the next chapter, we go further and look at the Agent SDK, which lets you build entirely custom AI-powered developer tools.

Chapter 15: The Claude Agent SDK

Up to this point, you've used Claude Code as a finished product — a general-purpose AI developer tool you interact with through your terminal or IDE. But what if you want to build something more specific? A code review bot that runs on every pull request. A documentation generator tailored to your company's format. An automated migration tool that understands your specific framework.

The Claude Agent SDK (formerly the Claude Code SDK) is what makes that possible. It's the same foundation Claude Code is built on, exposed as a library you can use to create your own AI-powered developer tools. Where Claude Code is the Swiss army knife, the Agent SDK lets you build a scalpel — purpose-built for the task you need.

This chapter covers what the SDK is, when to reach for it instead of Claude Code directly, and how to build real tools with it.

What Is the Agent SDK?

The Claude Agent SDK is a software development kit for building custom AI agents powered by Claude. An "agent" here is a program that can reason about a task, use tools to accomplish it, and make decisions along the way — the same loop that powers Claude Code itself.

The SDK gives you:

- **Conversation management**: send messages to Claude, receive responses, maintain context
- **Tool definition**: define custom tools that Claude can invoke during a conversation
- **Multi-turn orchestration**: manage the back-and-forth between Claude's reasoning and tool execution
- **Streaming**: receive Claude's output as it's generated, rather than waiting for the full response

- **Model configuration**: choose which Claude model to use, set parameters, manage costs

The SDK is available for Python and TypeScript. It handles the low-level details of API communication, tool call parsing, and conversation state management so you can focus on what your agent does.

Think of it this way: Claude Code is an agent built with the Agent SDK. It has specific tools (file read, file write, bash, search) and specific behaviors (ask for permission, follow CLAUDE.md). When you use the Agent SDK, you build your own agent with your own tools and your own behaviors.

When to Use the SDK vs. Claude Code Directly

This is the most important question in this chapter, and getting it wrong wastes significant time. The decision framework:

Use Claude Code directly when:

- You're doing interactive, one-off development work
- The task benefits from conversation and back-and-forth
- You need the full suite of built-in tools (file I/O, bash, search)
- The workflow is ad hoc and varies each time
- You're a human sitting at a terminal

Use the Agent SDK when:

- You're building an automated pipeline that runs without human interaction
- The task is well-defined and repeatable
- You need custom tools that don't exist in Claude Code
- You want to integrate AI reasoning into a larger application
- The tool will be used by other people or systems, not just you
- You need fine-grained control over the agent's behavior

The dividing line is automation and specialization. If you find yourself doing the same Claude Code workflow over and over — the same sequence of prompts, the same tools, the same checks — that's a candidate for an SDK-built tool. Encode the workflow once, and it runs automatically from then on.

SDK Basics

Start with the fundamentals. The Agent SDK follows a straightforward pattern: create an agent, give it tools, send it a task, and let it work.

Installation

For Python:

```
pip install claude-agent-sdk
```

For TypeScript:

```
npm install @anthropic-ai/claude-agent-sdk
```

A Minimal Agent

Here's the simplest possible agent in Python — one with no custom tools that just has a conversation with Claude:

```python
from claude_agent_sdk import Agent

agent = Agent(
    model="claude-sonnet-4-5",
    system_prompt="You are a helpful coding assistant.",
)

response = agent.run("What is the time complexity of quicksort?")
print(response.text)
```

This isn't very useful on its own — you could just use the Claude API directly. The SDK becomes powerful when you add tools.

Defining Tools

Tools are functions that your agent can invoke. You define them with a description (so Claude knows when to use them) and a handler (the code that runs when Claude invokes the tool).

```python
from claude_agent_sdk import Agent, tool

@tool(description="Read the contents of a file from the repository")
def read_file(path: str) -> str:
    """Read a file and return its contents."""
    with open(path, "r") as f:
        return f.read()

@tool(description="List files in a directory")
def list_directory(path: str) -> list[str]:
    """List all files in the given directory."""
    import os
    return os.listdir(path)

@tool(description="Run a shell command and return its output")
def run_command(command: str) -> str:
    """Execute a shell command."""
    import subprocess
    result = subprocess.run(
        command, shell=True, capture_output=True, text=True, timeout=30
    )
    return result.stdout + result.stderr

agent = Agent(
    model="claude-sonnet-4-5",
    system_prompt="You are a code review assistant. Use your tools to examine co
de.",
    tools=[read_file, list_directory, run_command],
)
```

Now your agent can read files, list directories, and run commands — a stripped-down version of what Claude Code does. Because you control the tools, you control exactly what the agent can and can't do.

The Agent Loop

When you call `agent.run()`, the SDK manages the agent loop automatically:

1. Your prompt is sent to Claude
2. Claude responds — either with text or a tool call
3. If it's a tool call, the SDK executes your tool handler and sends the result back to Claude
4. Claude continues reasoning with the tool result

5. Steps 2–4 repeat until Claude responds with final text (no more tool calls)

This loop is the core of agentic behavior. Claude doesn't just answer a question — it works through a problem step by step, gathering information and taking actions as needed.

You can configure the loop's behavior:

```
response = agent.run(
    "Review the code in src/auth.py",
    max_turns=20,           # Maximum number of tool-call rounds
    max_tokens=4096,        # Maximum tokens in Claude's response
    stop_on_error=False,    # Continue even if a tool call fails
)
```

Building a Code Review Bot

Let's build something real. A code review bot that analyzes pull request diffs and provides structured feedback. This is a textbook SDK use case: a well-defined, repeatable task that benefits from custom tools and runs automatically (triggered by CI).

Designing the Bot

Our code review bot needs to:

1. Fetch the diff for a pull request
2. Read the changed files
3. Analyze the changes for issues
4. Post review comments back to GitHub

We'll build custom tools for each of these steps.

The Tools

```
from claude_agent_sdk import Agent, tool
import subprocess
import json
import httpx
import os
```

```python
GITHUB_TOKEN = os.environ["GITHUB_TOKEN"]
REPO = os.environ["GITHUB_REPOSITORY"]  # e.g., "myorg/myrepo"

@tool(description="Get the diff for a pull request")
def get_pr_diff(pr_number: int) -> str:
    """Fetch the diff for a GitHub pull request."""
    response = httpx.get(
        f"https://api.github.com/repos/{REPO}/pulls/{pr_number}",
        headers={
            "Authorization": f"Bearer {GITHUB_TOKEN}",
            "Accept": "application/vnd.github.v3.diff",
        },
    )
    return response.text

@tool(description="Get the list of files changed in a pull request")
def get_changed_files(pr_number: int) -> list[dict]:
    """Get the list of changed files with their status."""
    response = httpx.get(
        f"https://api.github.com/repos/{REPO}/pulls/{pr_number}/files",
        headers={"Authorization": f"Bearer {GITHUB_TOKEN}"},
    )
    files = response.json()
    return [
        {"filename": f["filename"], "status": f["status"], "changes": f["changes
"]}
        for f in files
    ]

@tool(description="Read the current contents of a file from the repository")
def read_repo_file(path: str) -> str:
    """Read a file from the local repository clone."""
    with open(path, "r") as f:
        return f.read()

@tool(description="Run a linter or static analysis tool on a file")
def run_linter(file_path: str) -> str:
    """Run the project's linter on a specific file."""
    result = subprocess.run(
        ["npx", "eslint", "--format", "json", file_path],
        capture_output=True, text=True, timeout=30,
    )
    return result.stdout

@tool(description="Post a review comment on a pull request")
```

```python
def post_review(pr_number: int, body: str, event: str = "COMMENT") -> str:
    """Post a review on a GitHub pull request.

    Args:
        pr_number: The PR number
        body: The review body in Markdown
        event: APPROVE, REQUEST_CHANGES, or COMMENT
    """
    response = httpx.post(
        f"https://api.github.com/repos/{REPO}/pulls/{pr_number}/reviews",
        headers={"Authorization": f"Bearer {GITHUB_TOKEN}"},
        json={"body": body, "event": event},
    )
    return f"Review posted: {response.status_code}"
```

The Agent

```python
SYSTEM_PROMPT = """You are a thorough but pragmatic code reviewer. Your job is t
o
review pull requests and provide actionable feedback.

Guidelines:
- Focus on correctness, security, and maintainability
- Do not nitpick style issues (the linter handles those)
- Flag potential bugs, security vulnerabilities, and performance issues
- Acknowledge good patterns when you see them
- Be specific: reference file names, line numbers, and code snippets
- Categorize issues as: critical (must fix), suggestion (should consider),
  or nit (minor, optional)
- If the PR looks good, say so -- do not invent problems

Format your review as a structured Markdown document with sections for each
category of feedback."""

agent = Agent(
    model="claude-sonnet-4-5",
    system_prompt=SYSTEM_PROMPT,
    tools=[get_pr_diff, get_changed_files, read_repo_file, run_linter, post_revi
ew],
)

def review_pull_request(pr_number: int):
    """Run the code review bot on a pull request."""
    response = agent.run(
        f"Review pull request #{pr_number}. Start by getting the diff and "
```

```python
        f"changed files, read any files you need full context on, run the "
        f"linter on changed files, then post your review.",
        max_turns=30,
    )
    return response.text
```

Running It in CI

You can trigger this bot from a GitHub Actions workflow:

```yaml
name: AI Code Review
on:
  pull_request:
    types: [opened, synchronize]

jobs:
  review:
    runs-on: ubuntu-latest
    steps:
      - uses: actions/checkout@v4
        with:
          fetch-depth: 0

      - uses: actions/setup-python@v5
        with:
          python-version: "3.12"

      - run: pip install claude-agent-sdk httpx

      - name: Run AI Review
        env:
          ANTHROPIC_API_KEY: ${{ secrets.ANTHROPIC_API_KEY }}
          GITHUB_TOKEN: ${{ secrets.GITHUB_TOKEN }}
          GITHUB_REPOSITORY: ${{ github.repository }}
        run: python review_bot.py ${{ github.event.pull_request.number }}
```

This gives you automated AI code review on every pull request. The bot fetches the diff, reads changed files for context, runs the linter, and posts a structured review — no human intervention required.

Multi-Agent Workflows

Some tasks benefit from multiple specialized agents working together. The Agent SDK supports this through agent composition: one agent can invoke another as a tool.

The Pattern

Instead of building one agent that does everything, build several agents that each do one thing well, and a coordinator agent that delegates to them:

```
from claude_agent_sdk import Agent, tool

# Specialist agents
security_agent = Agent(
    model="claude-sonnet-4-5",
    system_prompt="You are a security specialist. Analyze code for vulnerabiliti
es.",
    tools=[read_repo_file, run_security_scanner],
)

performance_agent = Agent(
    model="claude-sonnet-4-5",
    system_prompt="You are a performance specialist. Identify performance issues.
",
    tools=[read_repo_file, run_profiler],
)

# Coordinator tools that invoke specialist agents
@tool(description="Run a security analysis on the changed files")
def security_review(file_list: str) -> str:
    """Have the security specialist review the given files."""
    response = security_agent.run(
        f"Analyze these files for security vulnerabilities: {file_list}"
    )
    return response.text

@tool(description="Run a performance analysis on the changed files")
def performance_review(file_list: str) -> str:
    """Have the performance specialist review the given files."""
    response = performance_agent.run(
        f"Analyze these files for performance issues: {file_list}"
    )
```

```python
    return response.text

# Coordinator agent
coordinator = Agent(
    model="claude-sonnet-4-5",
    system_prompt="""You coordinate code reviews. For each pull request:
    1. Get the changed files
    2. Send security-relevant files to the security reviewer
    3. Send performance-relevant files to the performance reviewer
    4. Synthesize findings into a single review""",
    tools=[get_changed_files, security_review, performance_review, post_review],
)
```

This pays off for complex workflows. Each agent has a focused system prompt and relevant tools. The coordinator orchestrates without needing to understand the details of each specialty.

When Multi-Agent Is Worth It

Multi-agent setups add complexity. They're worth it when:

- Different parts of the task require different expertise
- You want different models for different subtasks (a cheaper model for simple checks, a more capable model for nuanced analysis)
- The workflow has clear, independent phases that can run in parallel
- You need to keep each agent's context window focused

For simpler workflows, a single well-prompted agent with good tools is usually enough.

Streaming and Real-Time Output

For interactive tools, you want to show output as it's generated rather than waiting for the complete response. The SDK supports streaming:

```python
agent = Agent(
    model="claude-sonnet-4-5",
    system_prompt="You are a code explainer.",
    tools=[read_repo_file],
)
```

```
for event in agent.stream("Explain the architecture of this project"):
    if event.type == "text":
        print(event.text, end="", flush=True)
    elif event.type == "tool_use":
        print(f"\n[Using tool: {event.tool_name}]\n")
    elif event.type == "tool_result":
        print(f"[Tool completed]\n")
```

Streaming is essential for CLI tools where the user watches output in real time. It makes the tool feel responsive even when Claude is working through a complex multi-step task.

Error Handling and Resilience

Production agents need to handle failures gracefully. The SDK provides several mechanisms:

Tool Error Handling

When a tool raises an exception, the SDK catches it and sends the error message back to Claude as the tool result. Claude can then decide how to proceed — retry, try a different approach, or report the error.

```
@tool(description="Read a file from the repository")
def read_file(path: str) -> str:
    """Read a file, with clear error handling."""
    try:
        with open(path, "r") as f:
            return f.read()
    except FileNotFoundError:
        return f"Error: File '{path}' does not exist"
    except PermissionError:
        return f"Error: No permission to read '{path}'"
```

Returning error messages as strings (rather than raising exceptions) gives Claude the information it needs to adjust its approach. A `FileNotFoundError` exception becomes “the file doesn’t exist, so I should check the directory listing” in Claude’s reasoning.

Retry Logic

For transient failures (API timeouts, rate limits), wrap your agent calls with retry logic:

```
import time

def run_with_retry(agent, prompt, max_retries=3):
    for attempt in range(max_retries):
        try:
            return agent.run(prompt)
        except Exception as e:
            if attempt == max_retries - 1:
                raise
            time.sleep(2 ** attempt)  # Exponential backoff
```

Cost Controls

In automated pipelines, runaway costs are a real risk. A bug in your tool can cause Claude to loop indefinitely. Always set `max_turns` to a reasonable limit:

```
response = agent.run(
    "Review this pull request",
    max_turns=25,  # Hard limit on tool-call rounds
)
```

Monitor your API usage and set up billing alerts. An agent that runs on every pull request for a busy repository can consume significant tokens.

Testing Your Agents

Testing agents is more nuanced than testing regular functions because their behavior is non-deterministic. Some practical strategies:

Test tools independently. Each tool function is a regular function. Test it with known inputs and verify the outputs:

```
def test_get_changed_files():
    files = get_changed_files(pr_number=1)
    assert isinstance(files, list)
    assert all("filename" in f for f in files)
```

Test the agent end-to-end with known inputs. Create a test repository with known code and known issues, run the agent against it, and verify it identifies the expected problems:

```
def test_review_finds_sql_injection():
    response = review_agent.run(
        "Review the file test_fixtures/vulnerable.py"
    )
    assert "sql injection" in response.text.lower()
```

Use cheaper models for testing. During development, use a faster, cheaper model (like Claude Haiku) to iterate quickly. Switch to the full model for final validation.

Log tool calls. Record every tool invocation during test runs so you can debug unexpected behavior:

```
agent = Agent(
    model="claude-sonnet-4-5",
    system_prompt="...",
    tools=[read_file, run_linter],
    on_tool_call=lambda name, input: print(f"Tool: {name}({input})"),
)
```

Practical Tips

Start simple. Build a single-agent tool with two or three tools. Get it working end to end before adding complexity. Many useful tools are surprisingly simple.

Write good tool descriptions. Claude chooses tools based on their descriptions. Invest time in making them clear and specific. Bad descriptions lead to Claude using the wrong tool or not using tools when it should.

Keep system prompts focused. Your system prompt should tell the agent what it is, what it should do, and how it should behave. Don't overload it with every possible instruction — keep it focused on the specific task.

Monitor token usage. Agents can be token-hungry, especially with many tool calls. Log token counts for each run and watch for unexpected spikes.

Version your agents. Treat agent code like production code. Version-control it, review changes, and test before deploying. A broken agent in your CI pipeline blocks your entire team.

Summary

The Claude Agent SDK lets you build custom AI-powered tools that go beyond what Claude Code offers out of the box. It's the right choice when you need automation, specialization, or integration into larger systems.

The code review bot we built in this chapter is a template for many similar tools: fetch context, analyze it with Claude, take action on the results. The pattern applies to documentation generators, migration assistants, test writers, and any other well-defined development task.

In the next chapter, we shift from building tools to using tools — specifically, Claude Code's integrations with VS Code and JetBrains IDEs.

Chapter 16: IDE Integrations

There's a tension at the heart of Claude Code's design. It's a terminal-based tool, built for the command line, designed to be fast and keyboard-driven. But most developers spend the bulk of their time in an IDE — VS Code, IntelliJ, WebStorm, PyCharm. Switching between your editor and a terminal is friction, and friction slows you down.

Claude Code's IDE extensions resolve this tension. They bring Claude Code into your editor — not as a replacement for the terminal experience, but as a complement to it. You get the full power of Claude Code's agentic capabilities with the visual context of your IDE: file tree, open tabs, inline diagnostics, git decorations.

This chapter covers the VS Code and JetBrains extensions, how to set them up, how to use them effectively, and how to build a hybrid workflow that uses both the IDE extension and the standalone terminal.

The VS Code Extension

The Claude Code VS Code extension is the more mature of the two IDE integrations. It embeds a Claude Code session directly in your editor, with deep integration into VS Code's features.

Installation

Install the extension from the VS Code marketplace. Search for "Claude Code" (published by Anthropic) and click Install. Alternatively, install from the command line:

```
code --install-extension anthropic.claude-code
```

After installation, you'll see a Claude Code icon in the activity bar on the left side of VS Code. Clicking it opens the Claude Code panel.

The extension uses the same Claude Code installation you already have. If you've installed Claude Code globally via npm (`npm install -g @anthropic-`

`ai/claude-code`), the extension finds it automatically. If not, it prompts you to install it.

Authentication

The extension uses the same authentication as the CLI. If you've already authenticated Claude Code in your terminal (via API key or Claude subscription), the extension picks up those credentials. No separate login.

The Interface

The Claude Code panel in VS Code is a chat interface — you type messages, Claude responds, and the conversation flows like it does in the terminal. The IDE integration adds layers of context the standalone terminal doesn't have.

File context: When you have files open in your editor, Claude Code in VS Code is aware of them. You can reference the currently active file, and Claude Code can see your cursor position, your selection, and your open tabs. This makes requests like "refactor the function I'm looking at" work naturally.

Inline diffs: When Claude Code edits a file, the VS Code extension shows the changes as an inline diff — the same view you'd see in a code review. You can accept or reject individual changes without leaving your editor. This is a significant improvement over the terminal experience, where you see the raw edit output.

Terminal integration: The extension can run terminal commands in VS Code's integrated terminal, so you can see the output alongside your code. When Claude Code runs your tests, the results appear in a terminal pane below your editor.

Diagnostics awareness: VS Code's built-in diagnostics (errors, warnings from TypeScript, ESLint, etc.) are visible to Claude Code. Say "fix the errors in this file" and Claude Code sees the same red squiggles you see.

Key Features

Chat panel: The primary interface. Type your request, see Claude Code's response, approve or reject tool calls. The chat panel supports the same slash commands as the CLI: `/clear`, `/compact`, `/cost`, and so on.

Inline editing: Select code in your editor, then invoke Claude Code from the right-click context menu or a keyboard shortcut. Claude Code sees your selection and can operate on it directly — "explain this code," "refactor this function," "add error handling here."

Quick actions: The extension adds code actions (lightbulb menu items) for common operations. When you see an error, Claude Code may offer to fix it. When you select a function, it may offer to generate tests.

Status bar: The bottom of VS Code shows Claude Code's status — whether it's thinking, using a tool, or waiting for input. Token usage shows here too.

Keyboard Shortcuts in VS Code

Effective use of the VS Code extension depends on keyboard shortcuts. The defaults are reasonable, but customizing them to your preferences makes a real difference.

Default Keybindings

The extension ships with these default keybindings (which may vary slightly by platform):

- **Ctrl+Shift+P / Cmd+Shift+P** then type "Claude Code" — access all Claude Code commands from the command palette
- **Ctrl+L / Cmd+L** — focus the Claude Code chat panel (open it if closed)
- **Ctrl+I / Cmd+I** — inline edit mode (with selected code)
- **Ctrl+Shift+I / Cmd+Shift+I** — start a new Claude Code conversation

Customizing Keybindings

You can remap any Claude Code command through VS Code's standard keybinding system. Open the Keyboard Shortcuts editor (Ctrl+K Ctrl+S) and search for "Claude" to see every available command.

Some useful custom bindings to consider:

```
[
  {
```

```
    "key": "ctrl+shift+c",
    "command": "claude-code.focusChat"
  },
  {
    "key": "ctrl+shift+r",
    "command": "claude-code.reviewSelection"
  },
  {
    "key": "ctrl+shift+t",
    "command": "claude-code.generateTests"
  }
]
```

The goal is to make Claude Code reachable with a single keystroke from anywhere in your editor. You should never need to reach for the mouse to interact with Claude Code.

The JetBrains Extension

Claude Code also supports JetBrains IDEs — IntelliJ IDEA, WebStorm, PyCharm, GoLand, and the rest of the JetBrains family. The JetBrains extension provides similar functionality to the VS Code extension, though the integration points differ given the different IDE architecture.

Installation

Install the plugin from the JetBrains Marketplace. Go to Settings > Plugins > Marketplace, search for "Claude Code," and install it. Restart your IDE when prompted.

Like the VS Code extension, the JetBrains plugin uses your existing Claude Code installation and authentication.

The Interface

The JetBrains extension adds a tool window (typically on the right side of the IDE) with the Claude Code chat interface. The experience is similar to VS Code:

- Chat-based interaction with the same capabilities as the CLI
- Awareness of open files and editor context

- Inline diff display for code changes
- Terminal integration for command execution

JetBrains-Specific Features

Inspections integration: JetBrains IDEs have a strong inspection system (code analysis, warnings, suggestions). Claude Code can see these inspections and use them as context. "Fix all the warnings IntelliJ is showing" works because Claude Code can read the inspection results.

Refactoring awareness: When Claude Code suggests a rename or extract refactoring, the JetBrains plugin can use the IDE's built-in refactoring engine rather than doing a text-based find-and-replace. Rename refactorings update all references correctly, even across files.

Run configurations: Claude Code can interact with your JetBrains run configurations — running tests, starting servers, or executing build tasks through the IDE's run system rather than raw terminal commands.

Terminal-in-IDE vs. Standalone Terminal

You have three ways to use Claude Code:

1. **The IDE extension** (VS Code panel or JetBrains tool window)
2. **Claude Code in the IDE's integrated terminal** (running `claude` in VS Code's or JetBrains' terminal pane)
3. **Claude Code in a standalone terminal** (a separate terminal application)

Each has trade-offs, and understanding them helps you pick the right one for the situation.

The IDE Extension

Best for: Quick interactions, inline edits, code you're actively looking at. The extension sees your editor context (open files, selections, diagnostics) and provides visual diffs.

Drawbacks: The interface is more constrained than a full terminal. Long-running operations feel cramped in a side panel. Complex multi-step workflows can be harder to follow than in a terminal.

The Integrated Terminal

Best for: Longer sessions, complex workflows, tasks where you want the full terminal experience but also want to see your editor. Running `claude` in VS Code's terminal gives you the complete CLI experience with your code visible alongside.

Drawbacks: No inline diffs, no automatic file context from open tabs. You get the terminal experience, not the extension experience.

The Standalone Terminal

Best for: Deep focus sessions where you want Claude Code to have your full attention. Large refactoring tasks, debugging sessions, and complex multi-file operations often work better in a dedicated terminal where the output has room to breathe.

Drawbacks: No visual integration with your editor. You need to switch windows to see the code changes Claude Code makes.

The Practical Choice

Most experienced Claude Code users settle into a pattern: the IDE extension for quick, contextual interactions (explain this, fix this, test this) and the standalone terminal for bigger tasks (refactor this module, debug this issue, build this feature). The integrated terminal is a good middle ground when you don't know how large the task will be.

Using Claude Code Alongside GitHub Copilot

This question comes up constantly: should you use Claude Code or GitHub Copilot? For many developers, the answer is both — but for different things.

What Copilot Does Well

GitHub Copilot excels at inline code completion. As you type, it predicts what you're about to write and offers to complete it. This is a different interaction model from Claude Code. Copilot is reactive and instant — it works at the speed of typing. It's best for:

- Completing boilerplate code
- Writing repetitive patterns
- Implementing straightforward logic when you already know the shape of the solution
- Tab-completing function signatures, import statements, and common patterns

What Claude Code Does Well

Claude Code excels at reasoning about your codebase and executing multi-step tasks. It doesn't predict your next line — it plans and executes entire features. It's best for:

- Writing code that requires understanding multiple files
- Debugging complex issues
- Refactoring across a codebase
- Tasks that require running commands, checking results, and iterating
- Generating tests from existing code
- Any task where you describe what you want rather than typing it character by character

The Hybrid Workflow

With both tools active in VS Code, you get the best of both worlds:

1. **Start with Claude Code** for planning and scaffolding. Ask it to create the skeleton of a new feature — the file structure, the interfaces, the boilerplate.

2. **Switch to Copilot** for filling in implementation details. As you type the body of a function, Copilot completes the straightforward parts.

3. **Return to Claude Code** for review and iteration. Ask it to review what you have written, add error handling, or write tests.
4. **Use Copilot** for quick edits and one-line changes as you refine the code.

The key is recognizing which tool fits the current moment. If you know exactly what you want to type and just want it completed faster, that's Copilot. If you need to think through a problem, explore your codebase, or make coordinated changes, that's Claude Code.

Configuration

Both extensions can coexist in VS Code without conflict. Copilot handles inline completions; Claude Code handles the chat panel and agentic workflows. They don't interfere with each other.

Some developers disable Copilot's inline suggestions while actively using Claude Code for a large task to reduce visual noise. You can toggle Copilot with a keyboard shortcut or from the status bar.

Building a Hybrid IDE Workflow

Let's walk through a complete workflow that combines the IDE extension, keyboard shortcuts, and Claude Code's capabilities into a fluid development process.

The Setup

You're working on a web application in VS Code. You have the Claude Code extension installed, Copilot active for completions, and the following custom keybindings:

- **Cmd+L**: Focus Claude Code chat
- **Cmd+I**: Inline edit with Claude Code
- **Cmd+Shift+T**: Ask Claude Code to generate tests for the current file

The Workflow

Morning: Starting a New Feature

You open VS Code, press Cmd+L to focus Claude Code, and type:

> "I need to add a user notification preferences system. Users should be able to choose which types of notifications they receive (email, push, in-app) and how frequently. Show me a plan."

Claude Code examines your project structure, looks at your existing user model, checks your notification system, and proposes a plan: a new database table, an API endpoint, a settings component, and updates to the notification dispatcher.

You discuss the plan, make adjustments, and then say:

> "Go ahead and implement it. Start with the database migration and model."

Claude Code creates the files. You see inline diffs in your editor — the migration file, the model, the API route. You review each change, accept or reject, and move on.

Midday: Implementing Details

With the scaffolding in place, you open the notification preferences component and start typing the implementation. Copilot kicks in, completing React state hooks, form handlers, and JSX boilerplate. For the straightforward parts — mapping over notification types, rendering checkboxes — Copilot beats asking Claude Code.

When you hit a tricky part — implementing the frequency throttling logic — you select the function, press Cmd+I, and type:

> "Implement this function. It should throttle notifications based on the user's frequency preference: immediate, daily digest, or weekly digest."

Claude Code generates the implementation, shown as an inline diff. You review, accept, and move on.

Afternoon: Testing and Debugging

You press Cmd+Shift+T to generate tests for the new module. Claude Code reads the implementation, generates comprehensive tests covering normal cases, edge cases, and error conditions. The tests appear as a new file in your editor.

You run the tests. Three fail. You select the failing test output in the terminal, press Cmd+L, and say:

> "These tests are failing. Help me figure out why."

Claude Code reads the test output, examines the implementation, identifies a timezone bug in the daily digest logic, and fixes it. The inline diff shows exactly what changed. You accept the fix, run the tests again, and they pass.

End of Day: Review

Before committing, you ask Claude Code:

> "Review all the changes I've made today. Check for any issues I might have missed."

Claude Code reviews the day's work across all changed files, checking for consistency, error handling, edge cases, and alignment with the rest of the codebase. It flags two issues: a missing null check and an inconsistent error message format. You fix both.

This workflow — plan with Claude Code, scaffold with Claude Code, implement details with Copilot, debug and review with Claude Code — uses each tool where it's strongest.

Tips for IDE Integration

Keep the chat panel visible. Dock the Claude Code panel where you can see it while coding. The right side panel works well in VS Code — out of the way but visible. You want to glance at Claude Code's output without switching views.

Use the split editor. When Claude Code modifies a file, VS Code can show a split view with the original and modified versions side by side. This makes reviewing changes much easier than reading a raw diff.

Learn the keyboard shortcuts first. Before customizing anything, spend a week with the default keybindings. You'll discover which actions you perform most often and which shortcuts feel awkward. Then customize based on actual usage, not speculation.

Keep sessions short in the IDE. The IDE extension is best for focused, task-specific interactions. Start a new conversation for each distinct task rather than running one long session all day. This keeps context focused and reduces token usage.

Sync your settings. If you use Claude Code in both the IDE and the terminal, keep your CLAUDE.md files and settings.json consistent. Both contexts should have the same project instructions and the same MCP servers configured.

Use the IDE for visual tasks. When the task involves seeing code in context — reviewing diffs, understanding file structure, navigating references — the IDE extension beats the terminal. When the task is purely reasoning and execution, the terminal is often faster.

Troubleshooting Common IDE Issues

Extension not finding Claude Code: Make sure Claude Code is installed globally (`npm install -g @anthropic-ai/claude-code`) and that the `claude` command works in your terminal. The extension looks for the `claude` binary on your PATH.

Slow startup: The extension starts a Claude Code session when you first interact with it. If this takes more than a few seconds, check that your authentication is working and that any configured MCP servers can start. A failing MCP server can delay startup.

Inline diffs not showing: Make sure you haven't disabled VS Code's diff editor. Confirm the Claude Code extension is up to date and check that inline diff display is enabled in the extension settings.

Keybinding conflicts: Both Copilot and Claude Code register keybindings that may conflict with each other or with your custom bindings. Use the Keyboard Shortcuts editor's conflict detection (the warning icon) to identify and resolve them.

High memory usage: Running Claude Code as an extension alongside Copilot and your normal VS Code extensions can push memory usage up. If you hit slowness, check VS Code's Process Explorer (Help > Open Process Explorer) to identify memory hogs.

Summary

IDE integrations bring Claude Code into the environment where you spend most of your development time. The VS Code extension offers inline diffs, editor context awareness, and deep integration with VS Code's features. The JetBrains extension provides similar capabilities for IntelliJ-family IDEs.

The most effective workflow is hybrid: use the IDE extension for quick, contextual interactions and the standalone terminal for larger tasks. Use Claude Code alongside Copilot, leaning on each tool where it's strongest — Copilot for inline completions, Claude Code for reasoning and multi-step tasks.

The goal isn't to choose one tool or one interface. It's to build a workflow where the right tool is always within a keystroke's reach, and switching between them is effortless.

Part V: Domain-Specific Development

"Every great developer you know got there by solving problems they were unqualified to solve until they actually did it."

— Patrick McKenzie

Chapter 17: Frontend Development

You have a Figma mockup open in one tab, a blank React project in the other, and a deadline that somehow got closer while you were reading the requirements doc. Frontend development is where design meets engineering, and the gap between "this looks great in the mockup" and "this works in every browser on every screen size" is where most of the real work lives.

Claude Code is good at frontend work. It understands component hierarchies, CSS layout systems, accessibility requirements, and the state management patterns that hold modern web applications together. In this chapter, you'll learn how to use it for the full spectrum of frontend development — from generating individual components to building entire interfaces from design specifications.

Component Generation Across Frameworks

Whether you work in React, Vue, or Angular, the core workflow is the same: describe what you need, provide context about your existing patterns, and let Claude Code generate components that fit your codebase.

React Components

React is where Claude Code shines brightest on the frontend. The component model suits AI generation — components are self-contained units with clear inputs (props) and outputs (rendered UI).

Start with a clear description of what you need:

```
Build a UserProfileCard component that displays a user's avatar, name,
role, and a "View Profile" button. It should accept a user object as a
prop and follow the existing component patterns in src/components/.
```

Claude Code reads your existing components to understand your patterns — whether you use functional components with hooks, how you handle prop types (TypeScript interfaces vs. PropTypes), your naming conventions, and your file structure. If a `Button` component already exists, it imports and uses it rather than creating an inline button.

This is where your CLAUDE.md file pays dividends. If you've documented your component conventions — "all components use TypeScript, export as named exports, colocate styles in .module.css files" — Claude Code follows them automatically.

For more complex components, give Claude Code the full picture:

```
Build a DataTable component that supports:
- Sortable columns (click header to sort)
- Pagination (10/25/50 rows per page)
- Row selection with checkboxes
- A search/filter input
- Loading and empty states

Use our existing design tokens from src/styles/tokens.ts and follow
the compound component pattern like our existing Modal component.
```

Claude Code generates not just the main component but the supporting pieces — the header row, the pagination controls, the selection context — all wired together and following the patterns it found in your codebase.

Vue Components

With Vue, Claude Code adapts to whichever style your project uses. If your `.vue` files use the Composition API with `<script setup>`, that's what you get. If you're on an older codebase using the Options API, Claude Code matches that instead.

```
Create a NotificationBell component using our Composition API patterns.
It should show a bell icon with a badge count, and a dropdown panel
listing recent notifications. Use our existing useNotifications composable
for data.
```

Claude Code handles Vue-specific patterns well — computed properties, watchers, emits declarations, slot definitions, and the template syntax that trips up tools trained primarily on React.

Angular Components

Angular's more structured approach — decorators, modules, services, dependency injection — gives Claude Code more patterns to follow but also more boilerplate to generate correctly. The key is pointing it at your existing module structure:

```
Generate an Angular component for a file upload widget. It should
support drag-and-drop and file selection, show upload progress, and
emit events when uploads complete or fail. Register it in the
SharedModule. Follow the patterns in src/app/shared/components/.
```

Claude Code generates the component class, the template, the stylesheet, the spec file, and updates the module declaration — the full set of files Angular expects.

CSS and Styling

CSS is one of those areas where Claude Code saves outsized time relative to the complexity of the task. Not because CSS is hard to write, but because it's tedious to get right across every edge case.

Tailwind CSS

If your project uses Tailwind, Claude Code is fluent in utility classes. Rather than writing raw CSS, it generates the right combination of classes:

```
Style this card component using our Tailwind setup. It should have
a white background, subtle shadow, rounded corners, and a hover
effect that lifts it slightly. Match the spacing scale we use
elsewhere in the dashboard.
```

Claude Code reads your `tailwind.config.js` to understand your custom theme — color palette, spacing scale, breakpoints — and uses your custom utilities rather than defaults when they exist.

CSS Modules, Styled Components, and CSS-in-JS

Claude Code adapts to whatever styling approach your project uses. Find `.module.css` files? It generates CSS Modules. Find `styled-components` imports? It generates styled components. Tell it what you use, or better yet, let it figure it out from your existing code.

Design Token Integration

If you have a design token system — whether that is CSS custom properties, a JSON token file, or a TypeScript constants file — tell Claude Code about it:

```
We use design tokens defined in src/styles/tokens.ts. All colors,
spacing, typography, and shadows should reference these tokens. Never
use hardcoded values.
```

Put this in your CLAUDE.md and every component Claude Code generates will pull from your token system rather than hardcoding `#3B82F6` or `16px`.

Responsive Design

Getting a layout that works from a phone to a widescreen monitor is one of the most time-consuming parts of frontend development. Claude Code handles it well if you're specific about your breakpoint strategy.

```
Make the ProductGrid component responsive:
- Mobile (< 640px): single column, cards stack vertically
- Tablet (640px - 1024px): two columns
- Desktop (> 1024px): three or four columns depending on container width
Use CSS Grid and our existing breakpoint tokens.
```

Claude Code generates the responsive styles and — critically — thinks about what else changes at each breakpoint. It adjusts font sizes, hides secondary information on mobile, and converts horizontal layouts to vertical ones.

For more nuanced responsive work, you can have Claude Code analyze an existing page:

```
Review the checkout page at src/pages/Checkout.tsx and its styles.
It looks broken on mobile -- the form fields overflow the viewport
and the order summary sidebar overlaps the main content. Fix the
responsive layout.
```

Claude Code reads the component and its styles, identifies the layout issues, and applies fixes. It may convert a CSS Grid two-column layout to a single-column stack on mobile, adjust padding that assumed a wide viewport, or fix a flex container that wasn't wrapping properly.

Accessibility

This is where Claude Code provides value beyond saving time — it catches accessibility issues that most developers miss entirely. Claude Code knows the WCAG guidelines and applies them as it generates code.

When you ask Claude Code to build a component, it automatically:

- Adds appropriate ARIA attributes (`aria-label`, `aria-expanded`, `aria-describedby`)

- Ensures interactive elements are keyboard navigable
- Uses semantic HTML (`<nav>`, `<main>`, `<article>`) instead of generic `<div>` soup
- Provides proper focus management in modals and dropdowns
- Includes `alt` text prompts for images

But you can also use it as an accessibility auditor:

```
Review the navigation menu component at src/components/NavMenu.tsx
for accessibility issues. Check keyboard navigation, screen reader
support, ARIA attributes, color contrast, and focus indicators.
```

Claude Code reads the component and reports issues — a dropdown menu that traps focus, a button that uses `onClick` on a `<div>` instead of a `<button>`, color combinations that fail contrast ratios. Then you can ask it to fix each one.

For a broader sweep:

```
Scan all components in src/components/ for common accessibility
issues. Prioritize problems that would affect keyboard-only users
and screen reader users.
```

This kind of project-wide audit plays to Claude Code's strengths — it reads every component file and cross-references patterns across the codebase.

State Management

State management is where frontend architecture lives or dies, and where the gap between a junior and senior developer's approach is most visible. Claude Code can help you implement state management that scales.

Local State and Lifting State Up

For simpler cases, Claude Code helps you decide where state should live:

```
The FilterPanel and ProductList components both need access to the
current filter values. Right now FilterPanel manages the state
internally and there's no way for ProductList to know what filters
are active. Refactor this so both components can access the filter
state.
```

Claude Code reads both components, identifies the state that needs to be shared, and lifts it to the nearest common parent — or suggests a context if the components are far apart in the tree.

Redux, Zustand, and Other State Libraries

If your project uses a state management library, Claude Code generates the full pattern — store definition, actions, reducers, selectors, and the component connections:

```
Add a shopping cart feature using our Zustand store pattern.
I need: add to cart, remove from cart, update quantity, and
a computed total. Follow the patterns in src/stores/.
```

Claude Code reads your existing stores to match your conventions — how you structure slices, whether you use immer for immutable updates, your naming patterns for actions and selectors — and generates a new store that looks like your team wrote it.

Server State with React Query or SWR

For data fetching, Claude Code generates query hooks that handle loading states, error states, caching, and invalidation:

```
Create a custom hook useProducts that fetches products from our API,
supports pagination and filtering, and invalidates the cache when a
product is updated. Use React Query and follow the patterns in
src/hooks/queries/.
```

The generated hook includes proper TypeScript types, error handling, and cache key management — the boilerplate that's correct but tedious to write by hand.

Real-World Workflow: Building a Dashboard from a Design Mockup

Let's walk through a realistic scenario end to end. You've been handed a Figma mockup for an analytics dashboard. It has a header with navigation, a sidebar with filters, a main content area with several chart cards, and a data table below the charts. The design includes desktop, tablet, and mobile breakpoints.

Step 1: Analyze the Design and Plan the Component Tree

Start by describing the layout to Claude Code:

```
I'm building an analytics dashboard with this structure:
- Header: logo, main nav links, user menu dropdown
- Sidebar: date range picker, metric selector checkboxes,
  "Apply Filters" button
- Main content area:
  - Row of 4 KPI summary cards (revenue, users, orders, conversion rate)
  - Row of 2 chart cards (line chart for trends, bar chart for
    top products)
  - Data table showing recent orders with sorting and pagination

Plan the component tree and file structure for this. We use React
with TypeScript, Tailwind CSS, and our existing component library
in src/components/ui/.
```

Claude Code proposes a component hierarchy — the page component, the layout sections, and the individual widgets — along with a file structure that follows your project conventions. Review this plan before moving forward. It's much easier to adjust the architecture now than after fifty files have been generated.

Step 2: Build the Layout Shell

```
Build the DashboardLayout component first. It should render the
header, sidebar, and main content area using CSS Grid. The sidebar
should collapse to a hamburger menu on mobile. Don't fill in the
actual content yet -- just use placeholder divs with labels so I
can verify the layout works.
```

This gives you a working layout you can view in the browser before investing time in the individual components. Claude Code generates the grid layout, the responsive breakpoints, and the mobile sidebar toggle.

Step 3: Build Components Bottom-Up

Now build the leaf components — the ones with no dependencies on other new components:

```
Build the KPISummaryCard component. It displays a metric label,
the current value (formatted as currency, number, or percentage
depending on the type), the change from the previous period
```

```
(green arrow up or red arrow down), and a sparkline. Use the
existing Card component from our UI library as the wrapper.
```

Then the chart containers:

```
Build the TrendChart component that wraps a Recharts LineChart.
It should accept an array of data points with date and value,
a title, and handle loading/empty states. Use our chart color
palette from the design tokens.
```

Build each component, verify it renders correctly, then move on. Bottom-up means each component works independently before you wire them together.

Step 4: Wire Up State and Data Fetching

```
Create a useDashboardData hook that fetches all the dashboard data
from our API. It should accept the current filter values (date range,
selected metrics) and return the KPI summaries, trend data, top
products data, and recent orders. Use React Query with proper
loading and error states.

Then update the DashboardPage component to use this hook and pass
the data down to each widget component.
```

Claude Code generates the data hook, connects it to your API client, and wires the data flow through the component tree. It handles loading states for each section independently — the KPI cards can show data while the charts are still loading.

Step 5: Add Interactivity

```
Add the filter functionality:
- The sidebar's date range picker should update the dashboard data
  when changed
- The metric checkboxes should show/hide the corresponding KPI cards
- The data table should support column sorting (client-side) and
  pagination (server-side via API)
- Add a "Download CSV" button to the data table that exports the
  current filtered view
```

Claude Code handles the state management wiring — connecting the sidebar filters to the data fetching hook, managing the local sort state in the table, and implementing the CSV export.

Step 6: Polish Responsiveness and Accessibility

```
Review the entire dashboard for responsive design issues. Test
these scenarios:
- Desktop (1440px): full sidebar visible, 4 KPI cards in a row
- Tablet (768px): sidebar collapses, 2 KPI cards per row, charts
  stack vertically
- Mobile (375px): everything single column, simplified table view

Also do an accessibility pass -- make sure all interactive elements
are keyboard accessible, the charts have proper descriptions, and
the color coding for positive/negative changes doesn't rely on
color alone.
```

Claude Code reads through every component, identifies where the responsive styles need adjustment, and fixes accessibility issues. It may add `aria-label` attributes to icon buttons, ensure the sidebar toggle announces its state to screen readers, and add text indicators alongside the colored arrows in the KPI cards.

Step 7: Clean Up and Review

```
Review all the dashboard components for:
- Consistent code style and naming
- Proper TypeScript types (no 'any' types)
- Missing error boundaries
- Performance issues (unnecessary re-renders, missing memoization)
- Hardcoded values that should use design tokens
```

This final review pass catches the rough edges that accumulate during rapid development. Claude Code finds the one component where you forgot to use the design token for spacing, the data table that re-renders on every parent state change because it's missing `React.memo`, and the API error that silently swallows instead of showing the user a message.

Tips for Frontend Work with Claude Code

Share your design tokens early. Before generating any components, point Claude Code at your design system — tokens, base components, utilities. This prevents the tedious work of going back and replacing hardcoded values later.

Build components in isolation first. Generate each component independently, verify it works, then compose them. If you ask Claude Code to build an entire page in one shot, the result is harder to debug when something doesn't render correctly.

Use your browser's dev tools as feedback. When a component looks wrong, inspect it in the browser, note the specific CSS issue, and tell Claude Code exactly what's wrong. "The gap between the cards is 24px but should be 16px" beats "the spacing looks off."

Don't skip the accessibility pass. It's tempting to treat accessibility as a nice-to-have, but Claude Code makes it nearly effortless. Ask for an accessibility review on every component and you'll ship a better product with minimal extra effort.

Let Claude Code handle the responsive tedium. Writing media queries and testing breakpoints is one of the most time-consuming parts of frontend development. Claude Code generates responsive styles faster than you can write them by hand, and it's less likely to forget an edge case because it approaches the task systematically rather than by visual spot-checking.

Frontend development with Claude Code isn't about replacing your design sense or your understanding of user experience. It's about eliminating the translation layer between "I know what this should look like and how it should behave" and "I have working code that does it." You still make the decisions. Claude Code handles the implementation.

Chapter 18: Backend Development

Backend development is where Claude Code's ability to hold an entire system in its head pays off most. A frontend component is mostly self-contained — it takes props and renders UI. A backend feature touches routing, validation, business logic, database queries, authentication, error handling, and logging. Every layer has to agree with every other layer. Miss one and the whole thing falls apart at runtime.

Claude Code handles this multi-layer coordination naturally. It reads your existing routes to understand your URL patterns, checks your database schema to write correct queries, follows your error handling conventions, and generates tests that exercise the full stack. In this chapter, you'll learn how to use it for the core tasks of backend development — and we'll walk through building a complete REST API from an OpenAPI specification.

API Design and Implementation

Whether you're building REST APIs, GraphQL endpoints, or gRPC services, the workflow starts the same way: describe what the API should do and let Claude Code generate the implementation that fits your existing patterns.

REST API Endpoints

For a typical CRUD endpoint:

```
Create a REST API for managing blog posts. Endpoints:
- GET /api/posts (list with pagination, filtering by status and author)
- GET /api/posts/:id (single post with author details)
- POST /api/posts (create, requires authentication)
- PUT /api/posts/:id (update, only by author or admin)
- DELETE /api/posts/:id (soft delete, only by author or admin)

Follow the patterns in our existing user and comment controllers.
```

Claude Code reads your existing controllers, models, and middleware to understand how your application structures API endpoints. If your Express app uses a controller-service-repository pattern, it generates code in that pattern. If your Django app uses class-based views with serializers, it follows that convention instead.

The advantage here is consistency. When you write endpoints by hand, conventions drift over time — one endpoint returns errors in a different format, another forgets to validate input, a third skips the audit log. Claude Code reads your existing code and replicates the patterns exactly.

Input Validation

Validation is something every developer knows they should do thoroughly but often shortcuts under time pressure. Claude Code doesn't shortcut it:

```
Add request validation for the blog post endpoints. Validate:
- Title: required, string, 1-200 characters
- Body: required, string, max 50,000 characters
- Status: optional, must be "draft", "published", or "archived"
- Tags: optional, array of strings, max 10 tags, each max 50 chars

Use our existing validation middleware pattern with Zod schemas.
```

Claude Code generates the validation schemas and wires them into the middleware chain. It handles edge cases you may not think to validate — stripping leading/trailing whitespace, rejecting nested objects where you expect flat strings, ensuring arrays are actually arrays and not strings.

GraphQL

For GraphQL APIs, Claude Code generates the type definitions, resolvers, and data loaders:

```
Add a GraphQL type and resolvers for our Product model. Include:
- Query: products (with pagination and filtering), product (by ID)
- Mutation: createProduct, updateProduct, deleteProduct
- Nested resolver for product.reviews (with DataLoader for N+1 prevention)

Follow the patterns in src/graphql/types/user.ts.
```

The DataLoader pattern is worth highlighting. N+1 query problems are one of the most common performance issues in GraphQL APIs, and they're easy to introduce accidentally. Claude Code generates DataLoaders by default for nested resolvers because it understands the pattern from your existing code.

Database Schemas and Migrations

Database work is where a single mistake gets expensive. A bad migration on a production database isn't something you can casually roll back. Claude Code helps you get migrations right by understanding your existing schema and generating migrations that are safe to run.

Schema Design

When starting a new feature, describe the data model:

```
Design the database schema for a multi-tenant SaaS notification system.
Requirements:
- Each tenant has their own notification preferences
- Notifications can be email, SMS, or in-app
- Users can mark notifications as read/unread
- We need to support notification templates with variable substitution
- Must be efficient to query "all unread notifications for user X"

We use PostgreSQL with Prisma. Look at the existing schema in
prisma/schema.prisma for conventions.
```

Claude Code designs the tables, relationships, and indexes. It considers query patterns when placing indexes — if you need to query unread notifications by user frequently, it creates a composite index on `(user_id, read_at)` rather than letting you discover the performance problem later.

Migrations

For changes to existing schemas, Claude Code generates migrations that handle the transition safely:

```
Add a "teams" feature to our application. Users should be able to
belong to multiple teams, each team has a name and an owner.
Generate the Prisma migration for this. Make sure existing users
aren't broken -- they should be assigned to a "Default" team.
```

Claude Code generates the migration with the data backfill step included. Adding a required foreign key to an existing table with data requires either a default value or a data migration step — Claude Code knows this and handles it automatically.

For complex migrations, you can ask Claude Code to think through the risks:

```
Review this migration before I run it. What are the risks? Will it
lock any tables? Is there data that could violate the new constraints?
What's the rollback plan?
```

This kind of migration review is valuable even for experienced developers. Claude Code reads the migration, cross-references it against your current schema and data patterns, and flags potential issues — a table lock on a high-traffic table, a NOT NULL constraint that will fail if any rows have null values, or a foreign key that references a table in a different schema.

Query Optimization

Once your schema exists, Claude Code helps optimize the queries that run against it:

```
This query is slow in production (taking 3-5 seconds):

SELECT orders.*, users.name, products.title
FROM orders
JOIN users ON orders.user_id = users.id
JOIN order_items ON order_items.order_id = orders.id
JOIN products ON order_items.product_id = products.id
WHERE orders.created_at > NOW() - INTERVAL '30 days'
AND orders.status = 'completed'
ORDER BY orders.created_at DESC
LIMIT 50;

The orders table has 2 million rows. Suggest indexes and query
rewrites to improve performance.
```

Claude Code analyzes the query, identifies missing indexes, suggests rewrites (perhaps using a covering index or restructuring the joins), and explains the reasoning behind each suggestion.

Authentication and Authorization

Auth is one of the highest-stakes features in any application. Get it wrong and you have a security vulnerability. Get it right but make it hard to use and your team builds workarounds that undermine it.

Authentication Implementation

```
Implement JWT-based authentication for our Express API:
- POST /api/auth/register (email, password, name)
- POST /api/auth/login (email, password) -> returns access + refresh tokens
- POST /api/auth/refresh (refresh token) -> new access token
- POST /api/auth/logout (invalidate refresh token)

Use bcrypt for password hashing, store refresh tokens in the database,
access tokens should expire in 15 minutes. Follow security best
practices.
```

Claude Code generates the full auth flow with proper security measures — bcrypt with appropriate salt rounds, secure token generation, httpOnly cookie configuration for the refresh token, rate limiting on the login endpoint. It doesn't just write the happy path; it handles the edge cases that matter for security.

Authorization and Role-Based Access Control

```
Implement RBAC for our API. Roles: admin, editor, viewer.
- Admins can do everything
- Editors can create, read, and update resources but not delete
- Viewers can only read
- Users should only see resources in their own organization

Create the middleware and apply it to our existing routes. Check how
our routes are structured in src/routes/ and add authorization
consistently.
```

Claude Code reads every route file, adds the appropriate authorization middleware, and creates the permission-checking logic. The cross-cutting nature of this task — touching every route in the application — is exactly the kind of work where Claude Code's ability to read and modify many files in a session pays off.

Background Jobs and Async Processing

Many backend features need work to happen outside the request-response cycle — sending emails, processing uploads, generating reports, syncing with external services.

```
Set up a background job system using Bull and Redis for our Node.js app.
Create these job types:
```

```
- SendEmail: renders an email template and sends via SendGrid
- ProcessUpload: takes an uploaded CSV, validates rows, and imports
  into the database
- GenerateReport: queries data for a date range and creates a PDF

Include retry logic, dead letter handling, and a basic admin endpoint
to check job status. Follow the patterns in our existing codebase.
```

Claude Code generates the job definitions, the queue configuration, the worker setup, and the monitoring endpoint. It handles the concerns that are easy to forget — what happens when a job fails halfway through (idempotency), how to avoid processing the same job twice (deduplication), and how to detect stuck jobs.

Error Handling and Logging

Consistent error handling is the difference between an application that is debuggable in production and one that produces mysterious 500 errors with no useful information.

Structured Error Handling

```
Review our error handling across the entire API. Right now different
controllers handle errors differently -- some return { error: message },
some return { message: message }, some just send a status code with
no body. Standardize everything to use our AppError class and the
global error handler in src/middleware/errorHandler.ts.
```

Claude Code reads every controller and route handler, identifies inconsistent error handling, and refactors each one to use your standardized pattern. This is a tedious but critical refactoring task that may touch fifty files — exactly the kind of work that's painful by hand but straightforward for Claude Code.

Logging

```
Add structured logging throughout our API using our Winston logger.
Log:
- Every incoming request (method, path, user ID if authenticated)
- Every outgoing response (status code, duration)
- Database query durations over 100ms
- All errors with stack traces and request context
- Background job start, completion, and failure

Use correlation IDs to trace requests across services.
```

Claude Code adds logging statements throughout the codebase, following your existing logging patterns. The correlation ID pattern — generating a unique ID at the start of each request and passing it through to every log statement — is something Claude Code implements end to end because it sees the full request lifecycle across middleware, controllers, services, and database calls.

Real-World Workflow: Building a REST API from an OpenAPI Spec

Let's walk through a realistic scenario. Your team has designed a REST API for a project management tool. The API is defined in an OpenAPI 3.0 specification, and your job is to implement it. The spec defines 15 endpoints across four resource types: projects, tasks, comments, and users.

Step 1: Analyze the OpenAPI Spec

Start by having Claude Code read and summarize the specification:

```
Read the OpenAPI spec at docs/api-spec.yaml and give me an overview:
- What resources and endpoints are defined?
- What authentication is required?
- What are the request/response schemas?
- Are there any complex relationships between resources?
```

Claude Code parses the YAML file and gives you a clear summary. This is also a good time to ask it to flag potential issues in the spec — missing error responses, inconsistent naming, endpoints that probably need pagination but don't define it.

Step 2: Scaffold the Project Structure

```
Based on the OpenAPI spec, scaffold the project structure for a
Node.js Express API with TypeScript. Create:
- The directory structure (controllers, services, models, routes,
  middleware, types)
- TypeScript interfaces generated from the OpenAPI schemas
- The Express app with basic middleware (CORS, body parsing,
  error handling)
- The route registration matching the spec's paths
- Empty controller functions for each endpoint (just returning 501)

Use our standard project structure. Don't implement the business
logic yet -- I want to verify the structure first.
```

This gives you a running application where every endpoint exists and returns "Not Implemented." You can verify the route structure matches the spec before writing any business logic.

Step 3: Set Up the Database

```
Based on the OpenAPI schemas and the relationships between resources,
design the PostgreSQL database schema using Prisma. Include:
- All models with proper field types and constraints
- Relationships (projects have many tasks, tasks have many comments,
  users own projects and are assigned to tasks)
- Indexes for the query patterns implied by the API (list tasks by
  project, list comments by task, filter tasks by assignee)
- Timestamps (createdAt, updatedAt) on all models
- Soft delete support for projects and tasks

Generate the initial migration.
```

Claude Code designs the schema based on the API spec, creates the Prisma model definitions, and generates the migration. It infers relationship cardinality from the API endpoints — if there's a `GET /projects/:id/tasks` endpoint, it knows tasks belong to projects.

Step 4: Implement Endpoints by Resource

Now implement the business logic, one resource at a time:

```
Implement the Projects resource endpoints from the OpenAPI spec:
- GET /api/projects: list with pagination, optional status filter
- GET /api/projects/:id: single project with task count
- POST /api/projects: create (requires auth, user becomes owner)
- PUT /api/projects/:id: update (only owner or admin)
- DELETE /api/projects/:id: soft delete (only owner or admin)

Implement the full stack: controller -> service -> Prisma queries.
Include input validation matching the spec's schemas. Follow the
patterns established in the scaffold.
```

Repeat for tasks, comments, and users. Doing one resource at a time means you can review and test each before moving on. If Claude Code makes a structural decision you don't like on the first resource, correct it before the pattern propagates to the others.

Step 5: Add Authentication and Authorization

```
Implement the auth flow defined in the spec:
- Bearer token authentication using JWT
- Register and login endpoints
- Authorization rules:
  - Projects: only members can view, only owner can update/delete
  - Tasks: only project members can CRUD
  - Comments: anyone in the project can create, only author can
    edit/delete

Apply the auth middleware to all endpoints that require it according
to the spec's security definitions.
```

Claude Code reads the OpenAPI spec's security schemes and applies them to each endpoint as defined. It generates the auth middleware and the resource-level authorization checks.

Step 6: Generate Tests

```
Generate integration tests for all endpoints using our test setup
(Jest + Supertest). For each endpoint, test:
- Success cases with valid input
- Validation errors with invalid input
- Authentication (401 for unauthenticated requests)
- Authorization (403 for unauthorized requests)
- Not found (404 for missing resources)
- Edge cases (empty lists, pagination boundaries, concurrent updates)

Use factory functions to create test data. Set up and tear down
the test database for each test suite.
```

Claude Code generates a comprehensive test suite. Run the tests and use any failures as feedback:

```
These 3 tests are failing. Read the test output and fix the issues:
[paste test output]
```

Step 7: Add Production Concerns

```
Before we ship this, add:
- Request rate limiting (100 req/min per user, 1000 req/min per IP)
- Request/response logging with correlation IDs
- Health check endpoint at GET /health
- Graceful shutdown handling
```

```
- OpenAPI validation middleware (validate requests against the spec
  at runtime)
```

This final layer adds the production hardening that separates a prototype from a deployable service.

Tips for Backend Work with Claude Code

Start with the data model. The database schema constrains everything else. Get it right first, then build the API layer on top. Changing a schema after controllers and tests are written creates cascading changes.

Implement one resource completely before moving to the next. This lets you establish patterns — controller structure, service layer conventions, test organization — on a single resource and then replicate them. If you implement all controllers first, then all services, then all tests, inconsistencies are harder to spot.

Use your existing code as the style guide. The more existing code Claude Code can read, the better it matches your conventions. If you're starting a greenfield project, implement one endpoint by hand first, then let Claude Code follow that pattern for the rest.

Have Claude Code generate the boring parts. Input validation, error handling, logging, CRUD operations — these are correct-but-tedious tasks where Claude Code saves the most time. Spend your attention on the business logic and architectural decisions.

Always review generated SQL and migrations. Claude Code writes good database code, but the cost of a bad migration in production is high enough that human review is non-negotiable. Ask Claude Code to explain what a migration does and what risks it carries before you run it.

Backend development with Claude Code is about letting the tool handle the mechanical translation — from API spec to routes, from data model to queries, from requirements to validation rules — while you focus on the design decisions and business logic that require human judgment.

Chapter 19: DevOps and Infrastructure

There's a particular kind of frustration that comes from spending an hour debugging a Dockerfile only to discover the issue was a missing `--no-cache` flag on an `apt-get install`. Or from staring at a GitHub Actions workflow that passes locally but fails in CI because of an environment variable you forgot to set. Or from a Terraform plan that says it will "replace" your production database when you only meant to add a tag.

Infrastructure code is unforgiving. It runs in environments you can't easily inspect, fails with cryptic error messages, and the feedback loop — push, wait for CI, read the logs, fix, push again — is agonizingly slow. Claude Code shortens that loop. It writes infrastructure code that works on the first try more often than not, because it has seen thousands of Dockerfiles, CI pipelines, and Terraform configurations and knows the patterns that actually work.

In this chapter, you'll learn how to use Claude Code for the full range of DevOps and infrastructure tasks — from writing Dockerfiles to deploying applications on Kubernetes.

Dockerfiles and Container Configuration

Docker is one of Claude Code's strongest areas. Dockerfiles follow predictable patterns with well-known best practices, and most of the mistakes developers make are the same mistakes everyone makes.

Writing Dockerfiles

Start with your application and let Claude Code figure out the right container configuration:

```
Create a Dockerfile for our Node.js API. It's a TypeScript Express
app that builds with `npm run build` and starts with `npm start`.
It needs to connect to PostgreSQL and Redis at runtime. Optimize for
small image size and fast builds.
```

Claude Code generates a multi-stage Dockerfile — a build stage that installs all dependencies and compiles TypeScript, and a production stage that copies only the compiled output and production dependencies. It uses a slim base image, orders layers for cache efficiency (copying `package.json` and `package-lock.json` before the source), and sets up a non-root user for security.

For more complex scenarios:

```
Create a Dockerfile for our Python ML application. It needs:
- Python 3.11 with specific system libraries (libgl1, libglib2.0)
- A requirements.txt with numpy, pandas, scikit-learn, and torch
- The PyTorch install should use CPU-only to save image size
- The model weights file at models/classifier.pkl (800MB) should
  be in a separate layer so code changes don't re-download it
- Health check endpoint
```

Claude Code handles the nuances — ordering layers so the large model file is cached separately, using `--no-cache-dir` for pip to reduce image size, and configuring the health check correctly.

Docker Compose

For local development environments with multiple services:

```
Create a docker-compose.yml for local development. Services:
- Our API (build from local Dockerfile, hot reload with volume mount)
- PostgreSQL 15 with a persistent volume and init script
- Redis 7 for caching and job queues
- A worker service (same image as API, different command)
- Mailpit for email testing in development

Configure networking so services can reach each other, and expose
the right ports to the host. Add health checks so services start
in the right order.
```

Claude Code generates the compose file with proper `depends_on` conditions using health checks, named volumes for data persistence, environment variable configuration, and the volume mounts needed for hot reloading during development.

Debugging Container Issues

When something goes wrong in a container, Claude Code helps diagnose the problem:

```
Our Docker build is failing with this error:
[output from failed build]
```

```
Read the Dockerfile and figure out what's wrong.
```

Claude Code reads the Dockerfile, identifies the issue — maybe a dependency that needs a build tool not installed in the slim base image, or a `COPY` command referencing a file outside the build context — and fixes it.

CI/CD Pipelines

CI/CD configuration is one of the most tedious forms of programming. The YAML syntax is finicky, the available actions and plugins vary by platform, and testing changes requires pushing commits and waiting for the pipeline to run. Claude Code cuts most of that tedium.

GitHub Actions

```
Create a GitHub Actions CI pipeline for our TypeScript monorepo. It should:
- Trigger on push to main and on pull requests
- Run in a matrix: Node 18 and Node 20
- Install dependencies with npm ci (using a cache)
- Run linting (eslint)
- Run type checking (tsc --noEmit)
- Run unit tests with coverage
- Run integration tests against a PostgreSQL service container
- Upload coverage report to Codecov
- Block PR merge if any step fails
```

Claude Code generates the workflow YAML with the details right — the service container configuration for PostgreSQL, the caching strategy for `node_modules`, the matrix configuration, the correct action versions, and the proper ordering of steps.

For deployment workflows:

```
Add a deployment workflow that:
- Triggers when a tag matching v*.*.* is pushed
- Builds the Docker image and pushes it to our ECR registry
- Updates the ECS task definition with the new image
- Deploys to staging first, waits for health check
- Requires manual approval before deploying to production
- Notifies our Slack channel on success or failure
```

Claude Code generates the multi-environment deployment workflow with proper secret references, the manual approval gate using GitHub Environments, and the Slack notification step.

GitLab CI

If you use GitLab, the patterns are similar but the syntax differs:

```
Convert our GitHub Actions workflow to GitLab CI. We use GitLab's
shared runners. Keep the same stages and logic but use GitLab's
syntax -- stages, jobs, services, artifacts, and rules.
```

Claude Code handles the translation between platforms — GitHub's `services` become GitLab's `services`, action caching becomes GitLab's `cache` directive, and the approval gate uses GitLab's manual job feature.

Pipeline Optimization

Slow CI pipelines kill developer productivity. Claude Code can analyze and optimize them:

```
Our CI pipeline takes 18 minutes. Read the workflow file at
.github/workflows/ci.yml and suggest optimizations. The main
bottlenecks seem to be dependency installation and the integration
tests.
```

Claude Code reads the workflow and identifies optimizations — parallelizing independent steps, improving the caching strategy, using `--only-changed` flags for linting and testing, splitting the integration tests across multiple runners, or replacing a slow Docker build with a pre-built base image.

Infrastructure as Code

Terraform and similar tools bring the same version-control-and-review workflow to infrastructure that you use for application code. Claude Code writes infrastructure code that follows best practices and catches common mistakes before they hit `terraform apply`.

Terraform

```
Write Terraform configuration for our application's AWS infrastructure:
- VPC with public and private subnets across 2 AZs
```

```
- ECS Fargate cluster running our containerized API
- Application Load Balancer with HTTPS (ACM certificate)
- RDS PostgreSQL instance in the private subnet
- ElastiCache Redis cluster in the private subnet
- S3 bucket for file uploads with appropriate IAM policies
- CloudWatch log groups for all services

Use modules where appropriate. Follow Terraform best practices for
naming and organization.
```

Claude Code generates the Terraform files organized into logical modules — networking, compute, database, cache, storage. It sets up security groups with least-privilege rules, configures the RDS instance with encryption at rest and automated backups, and creates IAM roles with only the permissions each service needs.

For changes to existing infrastructure:

```
We need to add a CDN in front of our S3 bucket for serving user
uploads. Add a CloudFront distribution with:
- Origin access identity for the S3 bucket
- HTTPS only with our custom domain
- Cache policy for static assets (1 year max-age)
- Invalidation on deploy

Update the existing Terraform configuration. Don't break anything
that's already deployed.
```

That last instruction — "don't break anything" — is where Claude Code's ability to read and understand existing configuration matters. It reads your current Terraform state and configuration, understands what resources exist, and adds the CloudFront distribution with proper references to the existing S3 bucket rather than recreating it.

Reviewing Terraform Plans

Before applying changes, you can have Claude Code review the plan output:

```
Here's the output of terraform plan. Review it and tell me if anything
looks risky or unexpected:
[paste plan output]
```

Claude Code reads the plan and flags concerns — resources being replaced that you expected to be updated in-place, security groups with overly permissive rules, or changes that will cause downtime.

Kubernetes

Kubernetes configuration is verbose, repetitive, and easy to get subtly wrong. A missing resource limit, an incorrect readiness probe, or a misconfigured service selector causes problems that are hard to diagnose in a running cluster.

Manifests

```
Create Kubernetes manifests for our API deployment:
- Deployment with 3 replicas, rolling update strategy
- Resource requests and limits (the app typically uses 256Mi memory
  and 0.25 CPU)
- Readiness and liveness probes hitting /health
- ConfigMap for non-sensitive configuration
- Secret references for database credentials and API keys
- Service (ClusterIP) and Ingress with TLS
- HorizontalPodAutoscaler scaling between 3 and 10 replicas based
  on CPU utilization
```

Claude Code generates the manifests with proper label selectors, appropriate probe timing (giving the application time to start before the liveness probe kicks in), and a pod disruption budget that ensures zero-downtime deployments.

Helm Charts

For reusable deployments:

```
Convert our Kubernetes manifests into a Helm chart. Parameterize:
- Image repository and tag
- Replica count (min and max for HPA)
- Resource requests and limits
- Environment-specific configuration (dev, staging, production)
- Ingress host and TLS settings
- Database connection details

Create values files for dev, staging, and production environments.
```

Claude Code generates the chart structure with templates, helpers, and environment-specific values files. It parameterizes the right things — you don't want the health check path hardcoded in the template, but you probably do want the container port as a parameter.

Monitoring and Alerting

Infrastructure you can't observe is infrastructure you can't operate. Claude Code helps set up monitoring and alerting configurations.

```
Set up Prometheus monitoring for our Kubernetes deployment:
- ServiceMonitor to scrape our API's /metrics endpoint
- Grafana dashboard with panels for:
  - Request rate, error rate, and latency (RED metrics)
  - Pod CPU and memory utilization
  - Database connection pool usage
  - Background job queue depth
- Alerting rules:
  - Error rate above 1% for 5 minutes
  - P99 latency above 2 seconds for 10 minutes
  - Pod restarts in the last hour
  - Disk usage above 80%
```

Claude Code generates the Prometheus configuration, the Grafana dashboard JSON, and the alerting rules. The dashboard is the most valuable piece — building Grafana dashboards by hand through the UI is time-consuming, and doing it through JSON is tedious. Claude Code generates a well-organized dashboard with the panels you need.

Real-World Workflow: Containerizing and Deploying an Application

Let's walk through taking an application from "runs on my machine" to "deployed in production with CI/CD." The application is a Node.js API with a PostgreSQL database, currently running directly on a developer's laptop.

Step 1: Containerize the Application

```
Read our application code and create a production-ready Dockerfile.
The app is a TypeScript Express API in the src/ directory. It builds
with `npm run build`, the output goes to dist/, and it starts with
`node dist/server.js`. It needs these environment variables at
runtime: DATABASE_URL, REDIS_URL, JWT_SECRET, PORT.
```

Claude Code reads the application, understands its build process, and generates a multi-stage Dockerfile. It also creates a `.dockerignore` file to keep the image clean.

```
Now create a docker-compose.yml for local development that runs
the API, PostgreSQL, and Redis. The API should hot-reload when
source files change.
```

Test locally: `docker compose up` should start everything and the API should respond to requests.

Step 2: Set Up CI

```
Create a GitHub Actions CI workflow that:
- Runs on every push and PR
- Installs dependencies, lints, type-checks, and runs tests
- On push to main: builds the Docker image and pushes to GitHub
  Container Registry (ghcr.io)
- Tags the image with the git SHA and 'latest'
```

Claude Code generates the workflow. You push it, and now every commit runs through CI. The first time you set up a CI pipeline with Claude Code, you'll be surprised how few iterations it takes to get it passing — Claude Code knows the common gotchas (missing environment variables for tests, database service containers, caching strategies) and handles them upfront.

Step 3: Provision Infrastructure

```
Write Terraform configuration for deploying to AWS:
- ECS Fargate for the application (no servers to manage)
- RDS PostgreSQL (db.t3.micro for now, we can scale later)
- ElastiCache Redis (cache.t3.micro)
- ALB with HTTPS using our domain from Route53
- VPC with proper networking (public subnets for ALB, private
  subnets for everything else)
- Parameter Store entries for secrets

Organize into modules: networking, database, cache, application.
Use a tfvars file for environment-specific values.
```

Claude Code generates the Terraform configuration. Review it carefully — especially the networking, security groups, and IAM permissions.

```
Add a Terraform backend configuration using S3 and DynamoDB for
state locking. Also add a GitHub Actions workflow that runs
terraform plan on PRs and terraform apply on merge to main.
```

Now your infrastructure changes go through the same PR review process as your application code.

Step 4: Configure Deployment

```
Add a deployment job to our CI workflow. After the Docker image is
pushed on merge to main:
1. Update the ECS task definition with the new image tag
2. Deploy the new task definition to the ECS service
3. Wait for the deployment to stabilize (new tasks healthy, old
   tasks drained)
4. If the deployment fails, roll back automatically
5. Post the deployment status to our Slack channel
```

Claude Code generates the deployment workflow using AWS's ECS deployment mechanisms. It configures the deployment circuit breaker so failed deployments roll back automatically, and it uses the ECS wait-for-stable command to block until the deployment is complete.

Step 5: Add Monitoring

```
Add monitoring and observability:
- CloudWatch dashboard showing request count, error count, latency,
  CPU/memory utilization, and database connections
- CloudWatch alarms for: error rate > 5%, CPU > 80%, memory > 80%,
  healthy host count < 1
- ALB access logs to S3
- Application logs to CloudWatch Logs with a 30-day retention

Generate these as Terraform resources alongside our existing
infrastructure.
```

Claude Code adds the monitoring configuration to the Terraform modules and generates the CloudWatch dashboard definition.

Step 6: Create the Staging Environment

```
Duplicate our infrastructure for a staging environment. It should
be identical to production but with smaller instance sizes
(db.t3.micro, single AZ, fewer ECS tasks). Use Terraform workspaces
```

```
or separate tfvars files -- whichever approach is better for our
setup.

Update the deployment workflow to deploy to staging automatically
on merge to main, and require manual approval before deploying to
production.
```

Claude Code refactors the Terraform configuration to support multiple environments and updates the GitHub Actions workflow with the staging-then-production deployment pipeline.

Step 7: Document the Setup

```
Create a deployment runbook that covers:
- How to deploy (merge to main and the pipeline handles it)
- How to roll back (re-deploy the previous version)
- How to access logs
- How to SSH/exec into a running container
- How to run database migrations in production
- Common failure modes and how to diagnose them
```

Claude Code generates a practical runbook based on the actual infrastructure it just built. It knows the specific AWS console URLs, CLI commands, and log locations because it created the resources.

Headless Mode for Automation

Claude Code can run non-interactively using the `--print` flag (or its shorthand `-p`), which makes it suitable for CI/CD pipelines, automation scripts, and batch processing. In headless mode, Claude Code processes a single prompt, outputs the response, and exits — no interactive session, no waiting for user input.

This pairs well with `--dangerously-skip-permissions` in sandboxed CI environments, where no human is present to approve actions. You can pipe prompts in, capture structured output with `--output-format json`, and limit the scope with `--max-turns`.

Here's an example of headless mode in a GitHub Actions CI pipeline that auto-generates a PR summary:

```
name: PR Summary
on:
  pull_request:
```

```yaml
    types: [opened, synchronize]

jobs:
  summarize:
    runs-on: ubuntu-latest
    steps:
      - uses: actions/checkout@v4
        with:
          fetch-depth: 0

      - uses: actions/setup-node@v4
        with:
          node-version: '20'

      - run: npm install -g @anthropic-ai/claude-code

      - name: Generate PR summary
        env:
          ANTHROPIC_API_KEY: ${{ secrets.ANTHROPIC_API_KEY }}
        run: |
          claude -p "Review the changes in this PR (use git diff origin/main...H
EAD) and write a concise summary of what changed and why." \
            --dangerously-skip-permissions \
            --max-turns 10
```

You can also use headless mode for local automation — running Claude Code as part of a pre-commit hook, or a shell script that processes files in batch. Any task that doesn't need back-and-forth conversation is a good candidate.

Tips for DevOps Work with Claude Code

Always review infrastructure code before applying it. Claude Code writes good Terraform and Kubernetes configurations, but the blast radius of a mistake in infrastructure is much larger than in application code. Run `terraform plan` and read the output. Have Claude Code review the plan if it's complex.

Use Claude Code to translate between platforms. Moving from GitHub Actions to GitLab CI, Docker Compose to Kubernetes, or CloudFormation to Terraform are tedious translation tasks. Claude Code handles them well because the concepts map directly even though the syntax differs.

Start with the simplest working setup. Ask Claude Code for a basic Dockerfile and CI pipeline first. Get that working. Then add complexity — multi-stage builds, caching, deployment workflows, monitoring. Layering complexity onto a working foundation is much easier than debugging a complex setup that has never worked.

Use Claude Code for the YAML tedium. CI/CD pipelines, Kubernetes manifests, and Terraform configurations are verbose. Most of the content is boilerplate that follows well-known patterns. This is exactly the kind of code where Claude Code's generation is most reliable and saves the most time.

Feed error messages back to Claude Code. When a pipeline fails or a deployment errors out, paste the full error output to Claude Code. Infrastructure errors are often cryptic, and Claude Code has seen most of them before. It can diagnose "Error: InvalidParameterException: The container does not meet the minimum required memory limit" faster than you can search for it.

DevOps and infrastructure work is often treated as a necessary evil — something that has to be done but isn't the "real" work of building features. Claude Code doesn't eliminate the need to understand your infrastructure, but it does eliminate the hours spent on YAML syntax, debugging typos in pipeline configurations, and searching for the right Terraform resource arguments. You spend your time on architecture and operational decisions. Claude Code handles the implementation.

Chapter 20: Working with Existing Codebases

Starting a greenfield project is fun. You pick the framework, choose the patterns, set up the tooling, and write the first code. Everything is clean and intentional. But most professional software development isn't greenfield. Most of the time, you're joining a project that has been under active development for years, written by people who are no longer on the team, using patterns that made sense at the time but may not now.

Working with existing codebases is one of Claude Code's most valuable applications. The tool's ability to read and understand large amounts of code — tracing call chains across files, identifying patterns and conventions, mapping out architectures — turns what is normally a weeks-long onboarding process into something much shorter. In this chapter, you'll learn how to use Claude Code to understand unfamiliar code, identify technical debt, and modernize legacy systems incrementally.

Onboarding to a New Codebase

The first day on a new codebase is overwhelming. You have a repository with hundreds or thousands of files, no documentation (or outdated documentation), and a vague understanding of what the application does. The traditional approach is to read code, ask questions, and slowly build a mental model over days or weeks. Claude Code compresses that process.

The First Conversation

Start broad. Point Claude Code at the repository and ask it to give you the lay of the land:

```
I just joined this project and I'm onboarding. Give me a high-level
overview of this codebase:
- What kind of application is this?
- What languages, frameworks, and major libraries does it use?
- What's the directory structure and what lives where?
```

```
- How is it built, tested, and deployed?
- What external services does it depend on?
```

Claude Code reads the package files, configuration files, directory structure, and key source files to give you an architectural overview. This takes a few minutes and saves you the first day of poking around trying to figure out what you're looking at.

Mapping the Architecture

Once you have the overview, go deeper:

```
Map out the main architectural layers of this application. How does
a request flow from the user's browser to the database and back?
What are the key modules and how do they communicate?
```

Claude Code traces the request flow — from the route definitions through middleware, controllers, services, and database queries — and gives you a narrative description of how the pieces fit together. This is the mental model that normally takes a week of code reading to develop.

For specific areas:

```
Explain the authentication system. How do users log in? How are
sessions managed? Where is the auth middleware defined and what
routes does it protect?
```

```
Explain the data model. What are the main entities and how do they
relate to each other? Are there any unusual patterns in the database
schema?
```

```
How does the background job system work? What jobs exist, when are
they triggered, and how are failures handled?
```

Each of these questions would take an experienced developer several hours to answer by reading the code. Claude Code answers them in minutes because it reads dozens of files in rapid succession and synthesizes the information.

Understanding the Build and Deployment Pipeline

```
Walk me through the build and deployment process. What happens when
someone merges a PR to main? What CI checks run? How does the
application get deployed to production?
```

Claude Code reads the CI configuration files, deployment scripts, Dockerfiles, and infrastructure code to give you the full picture. It notes important details — the deployment requires a manual approval step, the staging environment uses a different database, the CI runs a subset of tests to keep build times under ten minutes.

Finding the Important Patterns

Every codebase has its own patterns — conventions that aren't documented anywhere but are followed throughout the code. Understanding these patterns is essential to writing code that fits:

```
What patterns and conventions does this codebase follow? Look at:
- How are API endpoints structured (controller/service/repository)?
- How is error handling done?
- What testing patterns are used?
- How is configuration managed?
- Are there any custom abstractions or base classes I should know about?
```

Claude Code reads enough of the codebase to identify the patterns. It may tell you that every service class extends a `BaseService` that provides logging and transaction management, or that the team uses a custom `Result` type for error handling instead of exceptions, or that integration tests use a specific factory pattern for test data.

Understanding Unfamiliar Code

Beyond the initial onboarding, Claude Code helps you understand specific pieces of code you encounter during your work. This is the "explain this to me" workflow, and it's one of the most common ways developers use Claude Code day to day.

Reading Complex Functions

```
Explain what this function does, step by step:
src/services/billing/invoice-calculator.ts, the calculateInvoice function.
Why is it so complex? What edge cases is it handling?
```

Claude Code reads the function, traces its logic, and explains it in plain language. More importantly, it explains the *why* — "this branch handles prorated charges for mid-month plan changes" or "this retry loop exists because the payment provider occasionally returns transient errors."

Tracing Dependencies

```
I need to change the User model. What parts of the codebase depend
on it? Show me every file that imports or references the User model
or the users table.
```

Claude Code searches the codebase and gives you a dependency map. This matters before making changes — you need to know the blast radius. It finds not just the direct imports but the indirect dependencies: the serializers that transform User objects, the tests that create User fixtures, the migration files that reference the users table.

Understanding "Why Was This Done This Way?"

Sometimes code looks wrong or unnecessarily complex, but there's a reason it was written that way. Claude Code can help you understand whether the complexity is accidental or intentional:

```
This function has a 200ms sleep in the middle of it. Why? Is this
intentional? Look at the git history and any comments to understand
the context.
```

Claude Code reads the function, checks the git blame for when the line was added, reads the commit message, and looks for related comments or issues. It may find that the sleep was added to work around a race condition in a third-party API, with a commit message that says "temporary fix for #342 — remove when provider fixes their webhook timing."

Identifying Technical Debt

Every codebase accumulates technical debt over time. The challenge isn't identifying that debt exists — everyone knows it does — but understanding which debt matters and what to do about it. Claude Code helps you assess technical debt systematically.

Code Quality Assessment

```
Assess the code quality of the src/services/ directory. Look for:
- Code duplication
- Functions that are too long or too complex
- Inconsistent patterns or conventions
- Missing error handling
```

```
- Outdated dependencies or deprecated API usage
- Missing or inadequate tests
```

Claude Code reads through the service layer and gives you a prioritized list of issues. It may report that three services duplicate the same pagination logic (which should be extracted to a shared utility), that the payment service has no error handling for network failures, and that the email service uses a deprecated SendGrid API version.

Architecture Assessment

```
Assess the overall architecture for potential problems:
- Are there circular dependencies between modules?
- Are there god classes or god modules that do too much?
- Is the separation of concerns clean or are there layers that
  reach across boundaries?
- Are there performance bottlenecks visible in the code
  (N+1 queries, missing indexes, synchronous calls that should
  be async)?
```

Claude Code maps the dependency graph, identifies modules with too many responsibilities, and finds architectural violations. It may find that the controller layer queries the database directly in some places (bypassing the service layer), or that three different modules implement their own caching logic instead of using the shared cache service.

Dependency Audit

```
Review our dependencies (package.json / Gemfile / requirements.txt):
- Which dependencies are significantly out of date?
- Are any dependencies deprecated or unmaintained?
- Are there dependencies we're barely using that could be removed?
- Are there known security vulnerabilities?
```

Claude Code reads the dependency files, checks usage across the codebase, and identifies packages imported in only one file (and possibly replaceable with a few lines of code) or packages abandoned by their maintainers.

Incremental Modernization

The worst thing you can do with a legacy codebase is attempt a full rewrite. The second worst is leave it exactly as it is forever. The right approach is almost always incremental modernization — improving the codebase piece by piece while keeping it functional throughout.

Establishing the Strangler Pattern

The strangler fig pattern — building new functionality in a modern way while gradually replacing old code — is the safest approach to modernization. Claude Code helps you set it up:

```
We want to gradually migrate our Express.js API from JavaScript to
TypeScript. The codebase has 120 JS files. Set up the TypeScript
configuration so we can have both JS and TS files coexisting, then
migrate the simplest files first -- the utility functions in
src/utils/ -- to establish the pattern.
```

Claude Code configures `tsconfig.json` with `allowJs: true` so existing JavaScript continues to work, then migrates the utility files to TypeScript — adding type annotations, replacing `require` with `import`, and fixing any type errors that surface. This gives you a working migration pattern that can be applied to the rest of the codebase file by file.

Refactoring Without Breaking Things

When modernizing, the critical constraint is that the application must keep working at every step. Claude Code helps you make safe changes by reading the tests and running them:

```
Refactor the UserService to use async/await instead of callbacks.
The service has 15 methods, all using the callback pattern. Make
sure all existing tests still pass after the refactoring.
```

Claude Code converts each method, updates the callers, and runs the tests after each change. If a test breaks, it reads the failure and fixes the issue before moving on.

Extracting Modules

Large, monolithic files are one of the most common forms of technical debt. Claude Code helps you break them apart:

```
The file src/services/order-service.js is 2,400 lines long and
handles orders, payments, shipping, and notifications. Break it
into separate service classes -- OrderService, PaymentService,
```

```
ShippingService, and NotificationService -- with clean interfaces
between them. Update all imports across the codebase.
```

Claude Code reads the monolithic file, identifies the logical groupings, extracts each group into its own file with a clear interface, and updates every file in the codebase that imported from the original module. It handles the tricky cases — methods that touch multiple concerns and need to be split or delegated.

Working with Legacy Code Safely

Legacy code has a specific technical meaning: code without tests. Working with legacy code safely means adding tests before making changes, so you have a safety net that catches regressions.

Adding Tests to Untested Code

```
The PaymentProcessor class has no tests. Before I modify it, I need
a test suite that captures its current behavior. Write tests for
every public method, covering the main paths and the error cases
I can infer from the code. Don't try to improve the code yet --
just document what it currently does through tests.
```

This is characterization testing — writing tests that pass against the current behavior, whether that behavior is correct or not. These tests become your safety net. If a later change breaks one, you know you changed the behavior (intentionally or not) and can decide whether that's acceptable.

Claude Code is particularly good at this because it reads the implementation to identify the branches and edge cases, then writes tests that exercise each path. It finds the error handling branches, the boundary conditions, and the configuration-dependent behavior you might miss.

Understanding and Documenting Implicit Contracts

Legacy code often has implicit contracts — assumptions that aren't documented anywhere but that other parts of the system depend on:

```
Read the ReportGenerator class and document its implicit contracts.
What does it assume about its inputs? What side effects does it have?
What does calling code depend on about its outputs? I want to
understand what I can and cannot change safely.
```

Claude Code reads the class, traces its callers, and documents the contracts: "callers assume the returned array is sorted by date descending," "this method writes to the audit_log table as a side effect," "the output format must match what the PDF exporter expects on line 340 of report-exporter.js."

Real-World Workflow: Onboarding to a 5-Year-Old Rails Application

Let's walk through a realistic onboarding scenario. You've just joined a team that maintains a Rails application originally built five years ago. The original team is gone. Documentation is sparse. The application works, mostly, but nobody fully understands it.

Step 1: Get the Lay of the Land

```
I'm onboarding to this Rails application. Give me a comprehensive
overview:
- What version of Rails is it running?
- What does the application do (read the routes and models)?
- What's the database schema look like?
- What gems does it depend on and are any of them outdated or
  deprecated?
- How is the test suite organized and what's the coverage like?
- Are there any configuration files or environment variables I
  should know about?
```

Claude Code reads the Gemfile, `config/routes.rb`, the models directory, `db/schema.rb`, the test directory, and the configuration files. It produces something like:

"This is a Rails 6.1 application for managing employee onboarding workflows. It has 34 models, the most important being Employee, OnboardingPlan, Task, and Department. The database is PostgreSQL with 47 tables. Several gems are significantly outdated — notably Devise (3 major versions behind) and Sidekiq (2 major versions behind). The test suite has 420 tests with roughly 60% line coverage, concentrated in the model layer. Controller and integration tests are sparse."

This overview in the first ten minutes gives you a foundation for everything that follows.

Step 2: Understand the Domain Model

```
Explain the core domain model of this application. What are the
main entities, how do they relate to each other, and what's the
lifecycle of the primary workflow (employee onboarding)?
```

Claude Code reads the models, their associations, validations, and callbacks to explain the domain. It traces the onboarding lifecycle — an employee is created, an onboarding plan is generated based on their department and role, tasks are assigned to various stakeholders, and the plan tracks completion.

Step 3: Map the Hot Paths

```
What are the most important user-facing features? Look at the routes
and controllers to identify the main workflows. Which parts of the
code handle the most critical functionality?
```

Claude Code identifies the core workflows — creating new employees and generating their onboarding plans, the task dashboard where managers track progress, the notification system that reminds people of overdue tasks, and the reporting dashboard. These are the parts you need to understand first because they're the parts you're most likely to work on.

Step 4: Identify the Landmines

Every old codebase has landmines — code that looks harmless but will blow up if you touch it wrong:

```
Look for potential problems and gotchas in this codebase:
- Are there any callbacks or observers that trigger non-obvious
  side effects?
- Is there any code that depends on specific data being present
  in the database?
- Are there any monkey patches or runtime modifications?
- Is there any code that calls external services synchronously
  in request handlers?
- Are there database queries that could be slow on large datasets?
```

Claude Code finds the `after_save` callback on the Employee model that triggers a synchronous API call to the payroll system (which times out occasionally and causes 500 errors), the initializer that monkey-patches the String class with a custom `to_phone` method, and the report query that does a full table scan on the tasks table with no index on `completed_at`.

Step 5: Assess Modernization Priorities

```
Based on everything you've read, what are the highest-priority
improvements for this codebase? Consider:
- Security risks (outdated dependencies, vulnerable patterns)
- Reliability risks (missing error handling, no retries)
- Performance risks (slow queries, synchronous external calls)
- Developer experience (outdated tooling, missing tests)
- Technical debt that will slow down future feature work

Rank them by impact and effort.
```

Claude Code gives you a prioritized list:

1. **Upgrade Devise** (high impact, medium effort) — the current version has known security vulnerabilities
2. **Move the payroll API call to a background job** (high impact, low effort) — it causes timeout errors in production
3. **Add an index on tasks.completed_at** (high impact, trivial effort) — the report page takes 30 seconds to load
4. **Add controller tests for the onboarding workflow** (medium impact, medium effort) — the most critical path has no test coverage
5. **Upgrade Rails to 7.x** (medium impact, high effort) — needed eventually but not urgent

Step 6: Start Contributing

Now you have enough context to start working. When you pick up your first task, use Claude Code as an ongoing reference:

```
I need to add a feature that lets managers delegate onboarding tasks
to other team members. Before I start, tell me:
- Where is the task assignment logic currently?
- What model changes would I need?
- What controllers and views would I need to modify?
- Are there any existing tests I should update?
```

Claude Code reads the relevant code and gives you a roadmap for the feature — which files to modify, what patterns to follow, and what tests to update. You're contributing effectively within days rather than weeks.

Tips for Working with Existing Codebases

Start every session with context. When you come back to an unfamiliar codebase after a break, start your Claude Code session with a brief orientation: "I'm working on the Rails onboarding app. Today I need to modify the task assignment feature." This gives Claude Code the context it needs to read the right files.

Ask "why" before "how." Before asking Claude Code to change something, ask why the code is written the way it is. There may be a good reason for that strange pattern, and understanding it keeps you from introducing a regression.

Use Claude Code to write characterization tests before refactoring. If you're about to change code that has no tests, spend ten minutes having Claude Code write tests that capture the current behavior. This investment pays for itself immediately when your refactoring accidentally changes something.

Document what you learn. As Claude Code explains the codebase to you, capture the important findings in a CLAUDE.md file or in the project's documentation. This helps both Claude Code in future sessions and the next developer who joins the team.

Be skeptical of "just rewrite it" impulses. Claude Code makes it tempting to rewrite large sections of code because it can generate replacements quickly. Resist this with legacy code. Incremental improvement with tests at every step is safer than wholesale replacement, no matter how fast the replacement can be generated.

Working with existing codebases is where the "senior developer pair-programmer" metaphor for Claude Code is most apt. A good senior developer joining a new team doesn't rewrite everything they see. They read the code, understand the context, identify the important problems, and make targeted improvements. Claude Code helps you do exactly that — at a pace that would be impossible on your own.

Part VI: Operations and Collaboration

"Alone we can do so little; together we can do so much."

— Helen Keller

Chapter 21: Security and Safety

You have handed an AI agent the keys to your file system, your terminal, and potentially your production infrastructure. That sentence should give you pause. Claude Code is powerful precisely because it operates with real system access — it reads your files, writes new ones, executes shell commands, and interacts with external services through MCP servers. Every one of those capabilities is a vector for mistakes, data leaks, or unintended consequences if you do not treat security as a first-class concern.

This chapter is not about fear. It is about informed confidence. Once you understand the permission model, the sandboxing boundaries, and the practical habits that keep your projects safe, you can push Claude Code hard without lying awake wondering what it did while you were reading its output.

The Permission Model

Claude Code ships with a permission system that acts as a gatekeeper between Claude's intentions and your machine. Every potentially destructive action — writing a file, running a shell command, making a network request — requires explicit approval the first time it occurs in a session, unless you have pre-approved it.

How Permissions Work

When Claude Code wants to perform an action it has not yet been granted permission for, it pauses and asks. You see the exact command or file operation it wants to perform, and you choose to allow or deny it. You can also allow an action for the remainder of the session or permanently via your settings.

The `/permissions` command lets you review and manage current permission grants at any time. This is the single most important command for security-conscious users. Run it periodically, especially after long sessions where you may have approved things in a hurry.

Permission Categories

Permissions break down into several categories:

- **File reads** — Generally low risk but still relevant when working with sensitive codebases or configuration files containing secrets.
- **File writes** — Medium to high risk. A misguided write can corrupt source files, overwrite configuration, or introduce vulnerabilities.
- **Shell commands** — The highest-risk category. A shell command can do anything your user account can do, from installing packages to deleting directories to making network requests.
- **MCP tool calls** — Risk varies by server. A filesystem MCP server carries different risk than a Slack integration or a database connector.

The Dangerously Skip Permissions Flag

The `--dangerously-skip-permissions` flag does exactly what its name suggests: it bypasses the permission system entirely. Claude Code will read, write, and execute without asking. The flag exists for CI/CD pipelines and automated workflows where no human is present to approve actions.

Never use this flag in interactive sessions. Never use it on a machine with access to production systems unless the environment is fully sandboxed. The name is deliberately alarming because the Anthropic team wants you to think twice — and then think a third time — before using it.

If you must use it in automation, ensure the environment has minimal permissions itself. Run it in a container with no network access to production, no credentials in environment variables, and no SSH keys that reach sensitive servers.

System Access and Sandboxing

Claude Code runs as a process on your machine with the same permissions as your user account. There is no hypervisor, no container, no virtual machine separating Claude Code from the rest of your system by default. This is deliberate — sandboxing would prevent Claude Code from doing the real work developers need it to do.

What Claude Code Can Access

By default, Claude Code can access:

- All files your user account can read or write
- All shell commands your user account can execute
- Network resources reachable from your machine
- Environment variables set in your shell profile
- Git credentials configured on your system
- API keys and tokens stored in your environment

What You Should Restrict

Not everything that is accessible should be accessible. Practical measures:

Use a dedicated workspace. Do not run Claude Code from your home directory or a directory containing unrelated projects. Start sessions in the specific project directory you are working on. Claude Code's file access scopes naturally to the working directory and its children, but shell commands can reach anywhere.

Scrub environment variables. If your shell profile exports production database credentials, AWS keys, or API tokens you do not want Claude Code to see, use a separate shell profile or a tool like `direnv` to scope secrets to specific directories.

Use `.gitignore` and `.claudeignore`. Claude Code respects `.gitignore` when deciding which files to index and reference. You can also create a `.claudeignore` file to exclude files or directories from Claude Code's awareness. It works just like `.gitignore` — place it in your project root and use the same glob pattern syntax to specify files and directories Claude Code should never read or include in its context. Put your `.env` files, credential stores, and private keys in these ignore lists.

Limit MCP server permissions. Each MCP server you connect extends Claude Code's reach. A database MCP server lets Claude Code run queries. A cloud provider MCP server might let it provision resources. Only connect MCP servers you need for the current task, and disconnect them when you are done.

Handling Sensitive Data

Every codebase has secrets. Database connection strings, API keys, authentication tokens, private certificates — these are the crown jewels attackers target and accidental exposure can compromise.

Secrets in Your Codebase

Claude Code reads your files to understand your project. If secrets are hardcoded in source files, Claude Code will see them. More importantly, those secrets become part of the context sent to the Claude API. While Anthropic's data handling policies protect your data, the principle of least exposure still applies.

Best practices:

- **Never hardcode secrets.** Use environment variables, secret managers, or encrypted configuration files. This is good practice regardless of AI tools.
- **Use `.env` files with `.gitignore`.** Keep secrets in `.env` files that are excluded from version control and from Claude Code's awareness.
- **Audit before starting a session.** If you are about to work on a project with Claude Code for the first time, do a quick search for hardcoded secrets: `grep -r "password\|secret\|api_key\|token" --include="*.py" --include="*.js" --include="*.ts"`.
- **Review CLAUDE.md files.** Since CLAUDE.md files are automatically loaded, never put secrets in them. Use them for instructions that reference environment variable names, not values.

Sensitive Business Logic

Some codebases contain proprietary algorithms, trade secrets, or pre-release features your organization considers highly confidential. Remember that code you work on with Claude Code is sent to the Claude API as context.

If your organization has strict data handling requirements, consult your security team about Anthropic's enterprise offerings, which include options for zero data retention and dedicated infrastructure.

Code Review for AI-Generated Code

Claude Code writes good code. It also writes code that looks good but contains subtle issues. The gap between "looks correct" and "is correct" is where bugs, vulnerabilities, and maintenance nightmares live.

Why AI Code Needs Extra Scrutiny

AI-generated code has specific patterns that warrant attention:

Plausible but incorrect logic. Claude Code excels at producing code that reads naturally and follows conventions. That makes bugs harder to spot — your brain pattern-matches on the familiar structure and skips the logical error hiding inside.

Outdated patterns. Claude Code's training data is extensive, but it may occasionally suggest patterns that were standard in an older version of a library but have been superseded or deprecated.

Security anti-patterns. Watch for SQL queries built with string concatenation, missing input validation, overly permissive CORS configurations, hardcoded credentials in generated code, and authentication shortcuts.

Dependency suggestions. When Claude Code suggests adding a dependency, verify the package exists, is actively maintained, and has no known vulnerabilities. AI models can hallucinate package names, and typosquatting is a real supply chain attack vector.

A Code Review Checklist for AI Output

Use this checklist when reviewing code that Claude Code has written:

1. **Does it do what was asked?** Read the code against the original requirement, not just the AI's explanation of what it does.
2. **Are inputs validated?** Check function parameters, API endpoints, user-facing forms.
3. **Are errors handled?** Look for missing try/catch blocks, unhandled promise rejections, ignored return values.
4. **Are there security implications?** SQL injection, XSS, path traversal, insecure deserialization, exposed secrets.

5. **Does it match project conventions?** Naming, file organization, error handling patterns, logging.
6. **Are tests included?** If you asked for tests, do they actually test meaningful behavior, or are they trivially passing?
7. **Are dependencies legitimate?** Verify package names and check for known vulnerabilities.
8. **Is the scope appropriate?** Did Claude Code change files or systems beyond what was needed?

Using Claude Code to Review Its Own Output

A useful technique is to ask Claude Code to review code it just wrote, with a specific security focus: "Review the code you just wrote for security vulnerabilities, paying special attention to input validation and authentication." This is no substitute for human review, but it catches issues that a fresh analytical pass surfaces.

Common Security Pitfalls

These are the mistakes we see most often from developers using Claude Code in production workflows.

Running Claude Code with Elevated Privileges

Never run Claude Code as root or with administrator privileges. If Claude Code needs to perform a privileged operation, have it generate the command and run it yourself with `sudo` after review.

Trusting Without Verifying in CI/CD

Using Claude Code in automated pipelines with `--dangerously-skip-permissions` and broad system access is a recipe for disaster. Always scope CI/CD environments to the minimum permissions required. Use read-only credentials where possible. Run in isolated containers.

Ignoring the Diff

When Claude Code modifies files, review the diff before committing. Use `git diff` liberally. Set up your workflow so Claude Code's changes land in a staging area where you can inspect them before they become permanent. The Git integration in Claude Code makes this natural — use it.

Over-Permissioning MCP Servers

Connecting a database MCP server with write access when you only need to read data. Connecting a cloud provider MCP server with admin credentials when you only need to list resources. Every MCP connection should follow the principle of least privilege.

Session Hygiene

Long sessions accumulate permissions and context. If you shift from a frontend feature to configuring infrastructure, start a new session. The permission grants from your frontend work are irrelevant to your infrastructure task and only widen the surface area for mistakes.

Enterprise Security Considerations

Organizations adopting Claude Code across teams face additional security challenges that go beyond individual developer practices.

Data Residency and Compliance

Enterprise deployments need to consider where code context is processed. Anthropic offers enterprise plans with specific data handling guarantees. If your organization operates under GDPR, HIPAA, SOC 2, or similar frameworks, make sure your Claude Code deployment aligns with your compliance requirements.

Access Control and Authentication

Claude Code authenticates via your Anthropic account or API key. In team settings, ensure each developer uses their own credentials. Shared API keys make it impossible to audit who did what and create a single point of compromise.

Audit Logging

For regulated environments, maintain logs of Claude Code sessions, particularly commands executed and files modified. Git history captures file changes naturally, but shell command execution may need additional logging depending on your compliance requirements.

Network Security

Claude Code communicates with the Claude API over HTTPS. If your organization uses network monitoring or data loss prevention tools, configure them to inspect or at least log this traffic. Some organizations may want to route Claude Code traffic through a proxy for additional control.

Policy and Training

Technical controls are only half the picture. Develop clear policies for Claude Code usage in your organization:

- Which projects are approved for use with AI coding tools?
- What data classification levels are acceptable to process through external AI APIs?
- What review processes are required for AI-generated code before it reaches production?
- Who is responsible for AI-generated code — the developer who prompted it, the reviewer who approved it, or both?

Train your team on these policies and on the practices outlined in this chapter. The best permission model in the world does not help if developers reflexively approve every request without reading it.

Building a Security-First Workflow

Bringing it all together, here is a workflow that balances Claude Code's power with responsible security practices:

1. **Start clean.** Begin each session in the correct project directory with a scoped environment.
2. **Review permissions.** Use `/permissions` to verify your permission state, especially at the start of sessions.
3. **Read before approving.** When Claude Code asks for permission, read the actual command or operation. Do not approve on autopilot.
4. **Review the diff.** Before committing any AI-generated changes, review the diff carefully.

5. **Test the output.** Run the test suite. Try the feature manually. Do not assume correctness.
6. **Rotate sessions.** Start new sessions when switching contexts or after extended work.
7. **Audit periodically.** Review your Git history to understand the cumulative impact of AI-assisted changes.

Security is not a one-time configuration — it is a practice. The developers who get the most value from Claude Code use it confidently because they have invested in understanding its boundaries and built habits that keep those boundaries intact.

In the next chapter, we turn to a different kind of risk management: keeping your Claude Code usage affordable with practical cost management strategies.

Chapter 22: Cost Management

Claude Code uses Claude API tokens under the hood, and tokens cost money. This is not a subscription where you pay a flat fee and use all you want — every prompt you send, every file Claude Code reads for context, and every response it generates consumes tokens that appear on your bill. For individual developers, that can range from a few dollars a day to significantly more during heavy sessions. For teams, the numbers multiply fast.

The good news is that Claude Code gives you meaningful control over costs. Choosing the right model, structuring your prompts efficiently, managing context windows, and adopting smart workflow habits can cut your spend without sacrificing productivity. This chapter gives you the practical knowledge to keep your Claude Code usage sustainable.

Understanding the Token Economy

Before you can manage costs, you need to understand what you are paying for. Claude Code's pricing revolves around tokens — the units large language models use to process text.

What Is a Token?

A token is roughly four characters of English text, or about three-quarters of a word. Code tends to tokenize differently than prose — a single line of Python might be 15–30 tokens depending on variable name lengths and syntax. A 500-line source file might be 3,000–8,000 tokens.

Input vs. Output Tokens

You are charged separately for input tokens (everything you send to Claude, including context) and output tokens (everything Claude generates in response). Output tokens are typically more expensive than input tokens. The distinction matters because Claude Code often sends large amounts of context — your project files, conversation history, CLAUDE.md content — as input with every request.

The Context Window Tax

Here is the insight that surprises most new Claude Code users: you are not just paying for the question you typed and the answer you received. You are paying for the entire context window that accompanies every interaction. This includes:

- Your current prompt
- The conversation history for the session
- Files Claude Code has read into context
- CLAUDE.md files loaded automatically
- Tool call results (command outputs, file contents)
- System prompts and instructions

As your session grows longer, the context window grows larger, and every subsequent interaction costs more because all that accumulated context is re-sent with each request. This is the biggest driver of unexpected costs.

The Model Lineup and Their Costs

Claude Code supports multiple Claude models, and your model choice is the most impactful cost lever you have.

Opus 4.7

The flagship model. Most capable, most expensive. Opus excels at complex reasoning, architectural decisions, large refactors, and nuanced code generation. It produces the highest-quality output, at a premium price.

Use Opus when: you are tackling complex problems, debugging subtle issues, designing system architecture, or working on critical code where quality justifies the cost.

Sonnet 4.6

The balanced option. Sonnet offers strong coding capabilities at a much lower price point than Opus. For most everyday development tasks — writing functions, fixing bugs, generating tests, refactoring code — Sonnet delivers results that are hard to distinguish from Opus.

Use Sonnet when: you are doing routine development work, writing straightforward features, generating tests, or working through a list of implementation tasks.

Haiku 4.5

The economy option. Haiku is the cheapest model and the fastest. It handles simple tasks well: formatting code, writing boilerplate, answering quick questions, generating simple scripts.

Use Haiku when: you need quick answers, simple code generation, formatting, or any task where speed matters more than depth.

Switching Models Mid-Session

You can switch models during a session using the `--model` flag or by configuring your default in settings. A productive pattern is to start complex tasks with Opus, then switch to Sonnet or Haiku for follow-up implementation and cleanup. There is no penalty for switching — each request uses the selected model's pricing.

Monitoring Your Costs

You cannot manage what you do not measure. Claude Code provides tools to track your spending.

The /cost Command

The `/cost` command displays your current session's token usage and estimated cost. Run it regularly — at minimum, check it at the start and end of each session, and before embarking on a large task.

The output breaks down input tokens, output tokens, and total estimated cost. Watch how these numbers grow as your session progresses. If the input token count is climbing fast, your context window is growing and each interaction is getting more expensive.

API Dashboard

Your Anthropic account dashboard provides detailed usage data including daily and monthly breakdowns, per-model costs, and usage trends. Check it weekly if you are an individual developer, daily if you are managing a team budget.

Setting Spending Limits

The Anthropic API supports spending limits at the account level. Set a monthly limit that matches your budget. This is your safety net — it prevents a runaway session or a forgotten automation from generating an unexpectedly large bill.

For teams, consider setting per-user or per-project limits through your API key management. Different developers on different projects have different cost profiles, and granular limits help you see where your budget is going.

Strategies for Reducing Token Consumption

With the cost model in hand, here are practical strategies for spending less without accomplishing less.

Keep Sessions Short

Long sessions are expensive sessions. As conversation history accumulates, every new interaction re-sends all previous context. A session with fifty back-and-forth exchanges costs much more per interaction than a fresh session.

Start new sessions when:

- You are switching to a different task or area of the codebase
- Your conversation history has grown long and the early context is no longer relevant
- You notice the `/cost` command showing rapidly increasing per-interaction costs

Use /compact Aggressively

The `/compact` command summarizes your conversation history into a shorter form, shrinking the context window. Use it when you have been working for a while and the early parts of your conversation are no longer relevant.

Think of `/compact` as a reset button for context costs. After compacting, your next interaction sends much less context and costs correspondingly less.

Write Precise Prompts

Vague prompts lead to large responses, follow-up clarifications, and wasted tokens. Compare:

Expensive: "Help me with the authentication system."

Efficient: "Add a JWT refresh token endpoint to `src/auth/routes.ts` that accepts a valid refresh token and returns a new access token. Follow the patterns in the existing login endpoint."

The second prompt gives Claude Code exactly what it needs to produce a targeted response. No clarification round-trips, no speculative suggestions, no tokens wasted on the wrong interpretation.

Scope File Context Carefully

When Claude Code reads files for context, those files consume input tokens on every subsequent interaction. Be deliberate about which files you ask it to examine. Instead of "look at the whole project structure," try "read `src/auth/routes.ts` and `src/auth/middleware.ts`."

Leverage CLAUDE.md Efficiently

Your CLAUDE.md files are loaded automatically and included in every interaction's context. Keep them concise. A 2,000-word CLAUDE.md adds roughly 2,600 tokens to every single request in every session. Multiply that by hundreds of interactions and verbose instructions add up.

Include only information Claude Code needs for every interaction: coding conventions, key architectural decisions, and critical project context. Move task-specific or reference-heavy content to separate files that Claude Code reads only when needed.

Choose Output Wisely

Claude Code's responses consume output tokens, which are more expensive than input tokens. If you need a quick answer, say so: "Brief answer: what's the conventional way to handle..." prevents a detailed essay when a one-liner would do.

Similarly, when asking for code, be specific about scope. "Write just the function, no explanation needed" saves output tokens compared to the default of generating code with accompanying commentary.

Model Selection Strategy

Choosing the right model for each task is the highest-leverage cost optimization you have. Here is a practical framework.

The Three-Tier Approach

Tier 1 — Opus for architecture and complexity. Use Opus when you need deep reasoning: system design, complex debugging, security-sensitive code, performance optimization, or any task where a wrong answer creates significant downstream costs.

Tier 2 — Sonnet for daily development. Make Sonnet your default. It handles feature implementation, test writing, refactoring, documentation, and most bug fixes with quality close to Opus at a fraction of the cost.

Tier 3 — Haiku for quick tasks. Use Haiku for code formatting, simple boilerplate, quick lookups, commit message generation, and any task that is more mechanical than creative.

The Cost-Quality Tradeoff in Practice

A common mistake is defaulting to Opus for everything because "it is the best." This is like taking a taxi to the mailbox. Opus is the best, but for straightforward tasks Sonnet produces equivalent results. Your budget goes further when you match model capability to task complexity.

Run an experiment: spend a week using Sonnet as your default, escalating to Opus only when Sonnet's output falls short. Track your costs with `/cost`. Most developers find Sonnet handles 70–80% of their work with no noticeable quality difference.

Claude Max: The Subscription Alternative

For heavy users, Anthropic offers two Claude Max subscription tiers: Max 5x at $100/month and Max 20x at $200/month. These plans run on usage quotas rather than per-token billing, so costs are predictable even during heavy weeks.

When Claude Max Makes Sense

Calculate your average monthly API spend. If you consistently spend more than the Claude Max price, the subscription is a straightforward win. Even if your average spend is lower, the subscription offers predictability — you know exactly what Claude Code costs each month, which simplifies budgeting and removes the mental overhead of optimizing every interaction.

When Pay-Per-Token Makes Sense

If you use Claude Code intermittently — a few sessions per week, or only during certain project phases — pay-per-token may be cheaper. The subscription pays off only with consistent, heavy usage.

Team Subscriptions

For teams, evaluate Claude Max on a per-developer basis. Your heaviest users may save significantly with subscriptions, while occasional users are better served by shared API token pools. A mixed approach — subscriptions for power users, API tokens for light users — often provides the best overall economics.

Team Budgeting

Managing Claude Code costs across a team introduces challenges beyond individual optimization.

Establishing a Budget Framework

Start with visibility. Track costs per developer, per project, and per month. Most teams are surprised by the variance — some developers use ten times the tokens of others, not because they are wasteful but because their tasks demand more AI interaction.

Set budgets that reflect reality, not wishful thinking. Use the first month of tracked data as a baseline, then optimize from there. Arbitrary budgets set without data create friction and push developers to work around the system.

Cost Allocation

Decide how to allocate Claude Code costs:

- **Per developer** — Simple but unfair if developers work on tasks of different complexity.
- **Per project** — Aligns costs with the work that generates them, enabling ROI analysis.
- **Shared pool** — Simple to manage but provides no accountability or optimization incentive.

Most teams find per-project allocation strikes the best balance. It lets you answer "Is Claude Code cost-effective for this project?" with actual data.

ROI Calculation

Claude Code costs money, but it also saves time. A rough framework for ROI:

1. Track developer hours saved per week using Claude Code (self-reported is fine for estimation).
2. Multiply by the fully loaded hourly cost of a developer.
3. Subtract Claude Code costs for the same period.
4. The result is your net savings (or net cost).

Most teams find Claude Code pays for itself many times over, but having the numbers lets you make the case concretely and identify situations where usage patterns need adjustment.

Communicating About Costs

Frame cost conversations around value, not restriction. "How can we get more value per dollar from Claude Code?" is a more productive question than "How can we spend less?" Share the optimization tips from this chapter with your team. Celebrate developers who find efficient workflows, not just those with the lowest bills.

Cost Management Checklist

Here is a quick reference for keeping your Claude Code costs under control:

- **Set a monthly spending limit** on your Anthropic account.
- **Default to Sonnet** and escalate to Opus only when needed.
- **Run `/cost` regularly** to stay aware of session costs.
- **Use `/compact`** when sessions grow long.
- **Start new sessions** when switching tasks.
- **Write precise prompts** that minimize round-trips.
- **Keep CLAUDE.md files concise** to reduce per-interaction context costs.
- **Evaluate Claude Max** if your monthly spend is consistently high.
- **Track team costs** per project for ROI analysis.
- **Review the API dashboard** weekly.

Cost management is not about being stingy — it is about being intentional. Every token you spend should contribute to productivity. The techniques in this chapter help you eliminate waste so your budget goes toward the work that matters.

Next, we explore how teams can build shared workflows around Claude Code, from onboarding new developers to establishing AI-assisted code review practices.

Chapter 23: Team Workflows

Claude Code is powerful for individual developers, but its impact multiplies when a team adopts it together. Shared conventions, consistent CLAUDE.md files, and coordinated workflows turn Claude Code from a personal productivity tool into a team-wide force multiplier. The gap between a team where each developer uses Claude Code in their own idiosyncratic way and a team with aligned practices is enormous — the latter gets compounding returns as everyone benefits from shared context and institutional knowledge.

This chapter covers the practical mechanics of making Claude Code work for teams: from shared configuration and onboarding to AI-assisted code review and measuring the impact on your engineering organization.

Shared CLAUDE.md: Your Team's AI Playbook

The CLAUDE.md file is the single most important coordination mechanism for teams using Claude Code. When every developer on your team starts a session and Claude Code loads the same project-level CLAUDE.md, everyone gets consistent behavior. The AI follows the same conventions, uses the same patterns, and produces code that looks like it came from one cohesive team — because in a sense it did.

What Belongs in a Shared CLAUDE.md

A team CLAUDE.md should include:

Project architecture. A brief description of the system's structure, key directories, and how components connect. Claude Code uses this to make informed decisions about where to put new code and how to integrate with existing systems.

Coding conventions. Naming patterns, file organization rules, preferred patterns (e.g., "Use custom hooks for shared state logic, not context providers"), and anti-patterns to avoid. Be specific — "follow our style guide" is useless; "function components with Props suffix on type names, e.g., `ButtonProps`" gives Claude Code actionable direction.

Common tasks. How to run tests, start the dev server, deploy, and perform other routine operations. This lets any developer ask Claude Code to run the test suite without specifying the command.

Dependencies and tools. Which package manager, framework versions, testing library, and other tooling the project uses. This keeps Claude Code from suggesting incompatible tools.

Review standards. What the team expects from code review, what automated checks must pass, and what patterns reviewers will flag.

Maintaining the Shared CLAUDE.md

Treat your CLAUDE.md as living documentation. It should be version-controlled alongside your code and updated as part of your normal development process. When someone adds a new convention or changes an architectural pattern, the CLAUDE.md should be updated in the same pull request.

Assign ownership. Someone on the team — a tech lead, a senior developer, or a rotating responsibility — should review CLAUDE.md changes to keep the file focused and accurate. Without ownership, it either rots into irrelevance or bloats into a novel that taxes tokens on every interaction.

Personal vs. Project CLAUDE.md

Developers can also have personal CLAUDE.md files that apply across all their projects. Encourage team members to keep personal preferences — editor keybindings, preferred explanation verbosity, individual workflow patterns — in their personal files, and reserve the project CLAUDE.md for team-wide concerns. The separation keeps the shared file lean while letting individuals customize their experience.

Onboarding with Claude Code

New team members face a steep learning curve: unfamiliar codebase, unknown conventions, tribal knowledge locked in senior developers' heads. Claude Code, armed with a good CLAUDE.md and access to the codebase, becomes an always-on onboarding companion.

Setting Up New Developers

Create an onboarding checklist that includes Claude Code setup:

1. **Install Claude Code** and authenticate with the team's preferred method (API key or Claude Max subscription).
2. **Configure the default model** to the team standard (usually Sonnet for cost efficiency).
3. **Verify CLAUDE.md loading.** Have the new developer start a session and ask Claude Code about project conventions. If the answers reflect the CLAUDE.md content, the setup is correct.
4. **Walk through a sample task.** Pair with the new developer for their first Claude Code session on a real task. Show them how the team uses prompts, handles permissions, and reviews output.

Using Claude Code for Codebase Exploration

Encourage new developers to use Claude Code as a codebase guide:

- "Explain the request lifecycle in this application, from the API gateway to the database."
- "What are the main modules and how do they depend on each other?"
- "Walk me through how a user login works, starting from the frontend form submission."

These questions would normally require a senior developer's time. Claude Code, with access to the codebase and the context from your CLAUDE.md, can answer them accurately at any hour. This does not replace human mentorship, but it sharply reduces the time senior developers spend on routine explanations.

Documenting Tribal Knowledge

Onboarding often surfaces tribal knowledge — the things everyone knows but nobody has written down. Use Claude Code sessions with new developers as a chance to capture it. When a new developer asks Claude Code something and gets an incomplete or incorrect answer because the information is not in the codebase or CLAUDE.md, that is a signal to document it.

AI-Assisted Code Review

Code review is where Claude Code's team impact is most immediately visible. AI-assisted review does not replace human reviewers — it catches routine issues before a human ever looks at the code.

Pre-Review with Claude Code

Before submitting a pull request for human review, developers can ask Claude Code to review their changes:

"Review the changes in this branch against main. Check for bugs, security issues, naming convention violations, and missing tests."

This pre-review catches the easy stuff: typos in variable names, missing error handling, inconsistent patterns, obvious logic errors. With those resolved before human review, the human reviewer can focus on higher-value concerns — architecture, design trade-offs, and business logic correctness.

Structured Review Prompts

Create a standard review prompt that your team uses consistently. Put it in your CLAUDE.md or in a shared document:

```
Review this code change for:
1. Correctness: Does it do what the PR description says?
2. Security: Any injection, authentication, or data exposure risks?
3. Performance: Any obvious bottlenecks or N+1 queries?
4. Conventions: Does it follow our project patterns?
5. Tests: Are there tests, and do they cover meaningful behavior?
6. Edge cases: What inputs or scenarios might break this?
```

When every developer runs the same review prompt, you get consistent quality standards across the team.

Review as a Learning Tool

For junior developers, AI-assisted review doubles as a teaching tool. Instead of just flagging issues, Claude Code explains why something is problematic and suggests alternatives. A junior developer who runs Claude Code review on every PR before submission learns faster because they get immediate, contextual feedback on their code.

Limitations of AI Review

Be clear with your team about what AI review cannot do:

- It cannot assess whether the feature solves the actual business problem.
- It may miss issues that require deep domain knowledge.
- It does not replace the relationship-building and knowledge-sharing that human code review provides.
- It can be wrong — AI review suggestions should be evaluated, not blindly accepted.

Human review remains essential. AI review is a filter that improves the quality of what humans review, not a replacement for the review process.

Pair Programming with Claude Code

Pair programming with Claude Code is a different dynamic from traditional pair programming, but it can be equally productive.

The Human-AI Pair

In traditional pairing, one developer drives (writes code) and one navigates (reviews and directs). With Claude Code, you navigate while Claude drives. You provide direction, context, and judgment. Claude Code provides implementation speed, pattern knowledge, and tireless attention to detail.

This works best when you:

- **Think out loud.** Describe your reasoning as you go. "I want to add a caching layer here because the database query is slow. Let's use Redis with a 5-minute TTL." Clear direction produces better code.
- **Review incrementally.** Do not let Claude Code write 500 lines before you look at them. Ask for small increments — one function, one component, one test at a time.
- **Challenge the output.** "Why did you use a Map instead of an Object here?" or "Would this approach scale to 10,000 concurrent users?" Treating Claude Code as a collaborator you push back on produces better results than accepting everything.

Real-Time Collaboration Patterns

Some productive patterns for human-AI pairing:

Sketch and fill. You write the function signatures and high-level structure, then ask Claude Code to implement the bodies. You keep architectural control while Claude Code handles implementation details.

Test-first development. You write the tests, Claude Code writes the implementation to make them pass. This enforces your specifications and catches misunderstandings early.

Parallel exploration. You work on one part of a feature while Claude Code works on another in a separate session. Periodically merge your work and resolve conflicts. This is less "pairing" and more "parallel development," but it is a productive team-of-two pattern.

Team Guidelines for Claude Code Usage

As your team adopts Claude Code, you need guidelines that balance productivity with quality, security, and cost.

What to Standardize

- **Default model.** Agree on a default (usually Sonnet) to manage costs and ensure consistent output quality.
- **CLAUDE.md practices.** Define what goes in the shared file, how it is updated, and who owns it.
- **Review requirements.** Specify when AI-assisted pre-review is expected, and what still requires human eyes.
- **Permission policies.** Which shell commands should developers approve cautiously? What MCP servers are sanctioned?
- **Cost awareness.** Set expectations for `/cost` monitoring and model selection.

What Not to Standardize

- **Prompting style.** Developers have different communication styles and that is fine. What matters is the output quality, not the exact phrasing of prompts.

- **Session management.** Some developers prefer long sessions with `/compact`; others prefer frequent fresh starts. Both can be cost-effective.
- **Personal CLAUDE.md.** Individual preferences and workflow customizations are personal.

Writing a Team Claude Code Policy

Draft a one-page policy that covers:

1. Approved use cases (development, testing, documentation, code review).
2. Prohibited use cases (processing customer PII, accessing production databases, any use violating your organization's data policies).
3. Security requirements (no secrets in CLAUDE.md, review all shell commands, scope permissions).
4. Cost expectations (default model, budget awareness, escalation path for unexpected costs).
5. Quality requirements (all AI-generated code must pass existing review processes).

Keep it short. A ten-page policy will not be read. A one-page policy that covers the essentials gets read.

Measuring the Impact of Claude Code

If you are leading a team's adoption of Claude Code, you will eventually need to answer: "Is this making us more productive?" Anecdotes are nice; data is better.

Quantitative Metrics

Track these metrics before and after Claude Code adoption:

- **Cycle time.** How long from first commit to merged PR? Claude Code typically reduces this.
- **PR throughput.** How many PRs does the team merge per week? Watch for increases.
- **Bug rate.** Does the bug count per release change? Monitor for both improvements (Claude Code catching issues) and regressions (Claude Code introducing issues).

- **Time to first commit.** For new team members, how quickly do they contribute meaningful code? Claude Code often accelerates onboarding.
- **API costs.** Track Claude Code costs alongside the productivity metrics to calculate ROI.

Qualitative Signals

Numbers do not capture everything. Also pay attention to:

- **Developer satisfaction.** Are people enjoying their work more? Do they feel more productive?
- **Code quality perception.** Do reviewers feel that PR quality has changed?
- **Knowledge distribution.** Are fewer questions going to senior developers because Claude Code answers them?
- **Task avoidance reduction.** Are developers tackling tedious tasks (writing tests, updating documentation, refactoring legacy code) that they previously avoided?

Running a Pilot

If your organization is cautious, run a pilot:

1. Select two similar teams or two comparable projects.
2. One adopts Claude Code with the practices from this chapter; the other continues current practices.
3. Run the pilot for 4–8 weeks.
4. Compare metrics and gather qualitative feedback.
5. Use the results to make an informed adoption decision.

This approach gives you real data from your own organization, which is more convincing than any case study from another company.

Scaling Claude Code Across an Organization

Once a team proves Claude Code's value, the question becomes: how do you scale it?

Center of Excellence

Establish a small group (2–4 people) who become Claude Code experts. They develop best practices, maintain shared CLAUDE.md templates, train other teams, and troubleshoot issues. It does not need to be a full-time role — it works well as a rotating responsibility or a guild-style community of practice.

Shared Resources

Build a library of:

- CLAUDE.md templates for different project types (Appendix B provides starting points).
- Prompt libraries for common tasks.
- Cost optimization guides specific to your tech stack.
- Security checklists tailored to your compliance requirements.

Gradual Rollout

Roll out to teams in waves, with the center of excellence supporting each wave. The first wave is early adopters who are enthusiastic and tolerant of rough edges. The second wave benefits from the first wave's experience. By the third wave, you have a well-established playbook.

Force-feeding Claude Code to reluctant teams creates resentment and poor adoption. Let the results speak for themselves and let adoption be pull-based — teams should want Claude Code because they see what it does for other teams.

The AI development landscape is evolving rapidly. In the next chapter, we look at how to stay current with Claude Code's evolution and build sustainable practices for a future where AI-assisted development is the norm.

Part VII: Reference

"Documentation is a love letter that you write to your future self."

— Damian Conway

Chapter 24: The Evolving Landscape

You are reading a book about a tool that did not exist two years ago, built on technology that was a research curiosity five years ago, operating in a market that barely existed a decade ago. The pace of change in AI-assisted development is unlike anything the software industry has seen. New models, new capabilities, new tools, and new workflows appear monthly. The skills and practices you have learned throughout this book will serve you well, but only if you pair them with a deliberate approach to staying current.

This final chapter is about building that adaptability. We will cover how to stay informed about Claude Code's evolution, how to evaluate updates and new features, where AI development tools are heading, which human skills become more valuable as AI handles more implementation, and how to build a sustainable practice that does not burn you out or leave you behind.

Staying Current with Claude Code

Claude Code receives frequent updates — new model versions, new features, bug fixes, behavioral changes. Keeping up does not require obsessive monitoring, but it does require a system.

Official Channels

Anthropic's blog and release notes. Major model releases (like the jump from Claude 3.5 to Claude 4, and the rapid 4.5/4.6/4.7 cadence since) are announced here with detailed capability descriptions. Bookmark the blog and check it weekly, or subscribe to their RSS feed or newsletter.

Claude Code's changelog. Smaller updates — new slash commands, permission model changes, bug fixes — appear in the changelog. Run `claude --version` periodically to see your current version and whether updates are available.

The Anthropic developer documentation. API changes, new parameters, pricing updates, and deprecation notices all live here. If you are managing costs or building automations, this is required reading.

Community Sources

Developer forums and communities. Anthropic's own forums, relevant subreddits, Hacker News discussions, and Discord communities are where practical knowledge lives. Working developers share tips, workarounds, and use cases that official documentation does not cover.

Open source projects. Claude Code's ecosystem includes MCP servers, editor integrations, and workflow tools built by the community. Watch the popular repositories for innovations that might improve your workflow.

Conference talks and podcasts. As AI-assisted development matures, conference talks increasingly cover practical workflows rather than theoretical possibilities. Watch them at 2x speed for actionable insights.

Building a Learning Habit

Set aside 30 minutes per week to scan these sources. Not an hour, not a full afternoon — 30 minutes. Skim the release notes, check for new MCP servers, read one community post in depth. Small, consistent investments compound over time and spare you the jarring experience of discovering six months of changes all at once.

Evaluating Updates and New Features

Not every new feature deserves your immediate attention. New model releases do not automatically mean you should switch. New capabilities do not always apply to your workflow. Here is how to evaluate changes.

The Three-Question Filter

When a new Claude Code feature or model drops, ask:

1. **Does this solve a problem I have?** If you are not doing multi-file refactors, a 50% improvement in multi-file refactoring does not matter to you right now.
2. **What is the cost?** New models may be more expensive. New features may require workflow changes. New integrations may introduce complexity.

3. **Can I test it safely?** Try new features on a non-critical project first. Run the new model on a task you previously completed with the current model so you can compare quality and cost.

Avoiding Shiny Object Syndrome

The AI tool space is noisy. Every week brings announcements about revolutionary new capabilities, game-changing integrations, and paradigm-shifting workflows. Most are incremental improvements dressed in marketing language. The developers who get the most done are not the ones who chase every new tool — they are the ones who master their current tools and adopt new ones only when the evidence is clear.

This does not mean conservatism to the point of stagnation. It means being deliberate. When you hear about a new feature, write it down. Come back to it in two weeks. If it still seems relevant and other developers report positive experiences, evaluate it. If the community has already forgotten it, skip it.

Model Transitions

When Anthropic releases a new model generation, here is a practical evaluation process:

1. **Read the benchmarks** but do not trust them blindly. Benchmarks measure specific capabilities that may not reflect your workload.
2. **Run your own comparison.** Take five representative tasks from your recent work. Run them on both the old and new model. Compare output quality, speed, and cost.
3. **Check pricing.** New models sometimes change the price-performance equation. A more capable model at the same price is a clear win. A more capable model at 3x the price needs more thought.
4. **Migrate gradually.** Switch your default model for routine work first. Keep the old model available for tasks where you have tuned your prompts to its behavior.

The Trajectory of AI Development Tools

Where is all this heading? Prediction is risky, but some trajectories are clear enough to prepare for.

More Autonomy, More Oversight

AI development tools are moving toward greater autonomy. Today, Claude Code asks permission before running commands and waits for your approval. Tomorrow's tools will handle more complex workflows with less human intervention. It is already happening with headless mode and CI/CD integrations.

Greater autonomy demands greater oversight. As AI handles more steps independently, the human role shifts from approving individual actions to setting constraints, reviewing outcomes, and auditing behavior. Think of it as moving from micromanagement to setting guardrails and reviewing results.

Deeper Integration

Claude Code today is a terminal application with IDE integrations. The trend is toward deeper integration across the development lifecycle: design tools, project management, deployment pipelines, monitoring systems, incident response. MCP is the mechanism enabling this — each new MCP server connects Claude Code to another system in your workflow.

Prepare by building fluency with MCP and by thinking about which parts of your workflow could benefit from AI integration. Developers who have already mapped their workflow and identified integration points will adopt new capabilities faster than those starting from scratch.

Specialization

General-purpose coding assistance is becoming a commodity. The next wave of value comes from specialization — AI tools that understand specific domains, frameworks, and codebases deeply. Custom MCP servers, detailed CLAUDE.md files, and project-specific configurations are early examples.

Invest in your CLAUDE.md files and project-specific tooling. The more context and domain knowledge you give Claude Code, the more it behaves like a team member who understands your project rather than a generic coding assistant.

Multi-Agent Workflows

The industry is moving toward workflows where multiple AI agents collaborate, each handling a different aspect of development. One agent might handle code generation, another runs tests, a third reviews security, a fourth manages deployment. Claude Code's ability to run in headless mode and integrate with CI/CD pipelines positions it well for this multi-agent future.

You do not need to build multi-agent systems today, but knowing this is the direction of travel helps you make architectural decisions that will adapt to that future rather than fight against it.

Essential Human Skills in an AI-Augmented World

As Claude Code handles more implementation work, certain human skills grow more valuable, not less. These are the skills that will define excellent developers in the age of AI assistance.

Systems Thinking

Understanding how components interact, how changes propagate through a system, and how to design architectures that are maintainable and scalable — these are skills AI augments but does not replace. Claude Code can implement a caching layer, but you need to decide whether a caching layer is the right solution, what to cache, and how cache invalidation affects the rest of the system.

Problem Decomposition

Breaking a vague requirement into concrete, implementable tasks directly multiplies Claude Code's effectiveness. The developer who can decompose "improve the user onboarding experience" into a sequence of specific, well-scoped tasks gets far better results from Claude Code than one who types the vague requirement and hopes for the best.

Judgment and Taste

Knowing what "good" looks like — in code quality, user experience, system design, and trade-off decisions — is inherently human. Claude Code can generate five different approaches to a problem. You need the judgment to pick the one that is right for your context, your users, and your team.

Communication

As AI handles more implementation, developers spend more time communicating: defining requirements, reviewing code, explaining decisions, coordinating with stakeholders. Clear communication becomes the bottleneck, not typing speed.

That includes communication with AI. The developers who write clear, precise prompts also write clear, precise tickets, documentation, and emails. Prompt engineering is a special case of good communication.

Domain Knowledge

AI tools know code. You know your business, your users, your constraints, and your market. This domain knowledge is what turns generic code into a product that solves real problems. It cannot be replaced by any model, no matter how capable.

Security and Risk Assessment

As code generation accelerates, the ability to assess risks — security vulnerabilities, operational concerns, compliance implications, architectural debt — becomes critical. More code produced faster means more potential risks to evaluate. Developers who can efficiently assess and mitigate risks will be in high demand.

Building Sustainable Practice

AI-assisted development can be exhilarating. The speed, the capability, the feeling of accomplishing in hours what used to take days — it is exciting. But excitement leads to unsustainable practices if you are not careful.

The Acceleration Trap

When you can produce code faster, the natural response is to produce more code. More features, more projects, more commitments. But the non-coding parts of software development — thinking, planning, reviewing, testing, maintaining — do not accelerate automatically. If you fill all your saved time with more coding, you create maintenance debt that eventually overwhelms you.

Use the time Claude Code saves for the work that benefits from human attention: thoughtful design, thorough review, user research, documentation, and the kind of deep thinking that produces better architecture rather than just more code.

Maintaining Your Craft

There is a concern among developers that relying on AI tools will erode their coding skills. The concern is legitimate if you use Claude Code as a black box — type a requirement, accept the output, move on. It is not a concern if you use Claude Code as a learning tool — read the code it generates, understand the patterns, ask why it chose specific approaches, and write code yourself regularly.

Keep your fundamental skills sharp. Write code without AI assistance periodically. Solve algorithmic problems. Read other people's code. The developers who thrive are those who can work with AI and without it.

Managing the Cognitive Load

AI-assisted development changes the cognitive demands on developers. You spend less time typing and more time reviewing, evaluating, and directing. It is a different kind of mental work — often more intense because the decisions come faster. You review a hundred lines of generated code, decide it is correct, move to the next task, review another hundred lines.

Take breaks. Your ability to catch bugs in generated code degrades with fatigue like any other cognitive task. A five-minute break between intensive Claude Code sessions can prevent the mistakes that cost hours to fix later.

Setting Boundaries

Claude Code is available 24/7. You are not. The temptation to "just ask Claude Code to implement one more feature" at 11 PM is strong when it takes five minutes instead of an hour. But sustainable productivity requires sustainable hours. Use Claude Code to work smarter during your normal working hours, not to extend those hours into every waking moment.

Looking Forward

The landscape of AI-assisted development will look different in a year than it does today, and radically different in five years. But some principles will endure:

Tools change; fundamentals do not. The specific commands and features in this book will evolve. The principles — clear communication, deliberate practice, security awareness, cost consciousness — will remain relevant regardless of which tools you use.

AI is a multiplier. It multiplies good habits and bad habits equally. Invest in the habits worth multiplying.

The human element is not optional. AI does not remove the need for human judgment, creativity, and responsibility. It raises the bar for what we do with those capabilities.

Learning never stops. The developers who thrive in this landscape stay curious, experiment deliberately, and adapt continuously.

Epilogue

You started this book by installing a command-line tool. You are ending it as a developer who knows how to direct an AI agent through complex, real-world development work — from generating code and debugging failures to managing infrastructure, coordinating multi-file refactors, and building custom automation with MCP servers and hooks.

That is not a small shift. Terminal-native AI development changes the texture of your workday. The tedious parts — the boilerplate, the YAML, the mechanical refactors across dozens of files — compress into minutes. The interesting parts — architecture, design decisions, code review, debugging subtle logic — get more of your attention. The tool does not replace you. It frees you to do the work that requires your judgment.

The specific features and interfaces will evolve. Claude Code today is not what it will be a year from now. Models will get faster and more capable. New tools and integrations will appear. Pricing will change, slash commands might change, and capabilities will expand in ways that are hard to predict.

But the principles you have learned are durable. Context matters — a well-written CLAUDE.md file will be as valuable with future models as it is today. Verify the output — reviewing AI-generated code before committing it will never stop being important. Iterate — the conversational loop of prompt, review, refine works because it mirrors how humans collaborate, and that does not change with better models. Scope your tasks — clear, bounded requests produce better results from any tool, AI or otherwise.

If you came to this book from *Claude: The Definitive Guide* or *ChatGPT: The Definitive Guide*, you now have the full picture of how these AI tools fit into professional work. Claude Code is where the conversational AI capabilities you learned about in those books meet the realities of building and shipping software. If you have not read them, they are worth your time — the broader understanding of how these models think and where they struggle will make you a better Claude Code user.

Now close the book and open your terminal. You have code to write.

For quick reference on Claude Code commands, flags, and configuration, continue to the appendices.

Appendix A: Command Reference

This appendix is a reference for Claude Code's slash commands, CLI flags, configuration settings, and environment variables. Keep it handy during sessions — it is faster than asking Claude Code how to use itself.

Slash Commands

Slash commands are entered during an interactive Claude Code session. Type the command at the prompt and press Enter.

/help

Displays a summary of available slash commands and basic usage information. Your in-session quick reference when you cannot remember a specific command.

/clear

Clears the current conversation history and resets the context window. Use this when you want to start fresh without exiting and re-entering Claude Code. Note that it does not reset file permissions you have granted during the session.

/compact

Summarizes the current conversation history into a condensed form, shrinking the context window. Your primary tool for managing costs in long sessions. The conversation continues with the summarized context, so Claude Code retains the gist of what you discussed without carrying the full token weight of every previous exchange.

You can optionally provide instructions to guide the summarization: `/compact focus on the authentication changes we discussed`.

/cost

Displays token usage and estimated cost for the current session. Shows input tokens, output tokens, and total cost broken down by interaction. Run it regularly to stay aware of your spending. See Chapter 22 for cost management strategies.

/status

Shows the current session status including the active model, loaded CLAUDE.md files, connected MCP servers, and permission state. Useful for verifying your configuration, especially at the start of a session or after changing settings.

/init

Initializes a CLAUDE.md file in the current project directory. Claude Code examines your project structure and generates an initial CLAUDE.md with project context, inferred conventions, and common tasks. Treat the result as a starting point — review and customize it.

You can use `/init` in two ways: from the shell as `claude -p "/init"` (which uses non-interactive print mode to run the command and exit), or as `/init` typed at the prompt during an interactive session. Both produce the same result.

/exit

Ends the current Claude Code session and returns to your shell. Equivalent to pressing Ctrl+C or Ctrl+D. Useful when you want to close a session explicitly rather than via a keyboard shortcut.

/config

Opens or displays Claude Code configuration settings. Use this to view or modify settings like the default model, permission preferences, and other session-level configurations.

/permissions

Displays and manages permission grants for the current session. Shows which file operations and shell commands have been approved. Use this to audit your permission state and revoke permissions you no longer need.

/doctor

Runs diagnostic checks on your Claude Code installation. Verifies API connectivity, authentication, configuration file validity, and MCP server connections. Run it first when troubleshooting any issue — it catches the most common problems.

/login

Initiates or refreshes authentication with your Anthropic account. Use this if your session has expired or if you need to switch accounts.

/logout

Ends your current authentication session.

/model

Displays the current model or switches to a different one. Usage: `/model sonnet` or `/model opus` or `/model haiku`. See Chapter 22 for guidance on when to use each model.

/mcp

Displays information about connected MCP servers, including available tools and connection status. Use this to verify that your MCP integrations are working correctly.

/review

Asks Claude Code to review recent changes, typically the current git diff. A shortcut for the common workflow of requesting a code review before committing.

/terminal-setup

Configures your terminal for optimal Claude Code usage, including shell integration features.

CLI Flags

These flags are specified when launching Claude Code from the command line.

–model, -m

Specifies the Claude model to use for the session.

```
claude --model opus
claude -m haiku
claude --model sonnet
```

–print, -p

Runs Claude Code in non-interactive (print) mode. Sends a single prompt, prints the response, exits. Useful for scripting and one-off queries.

```
claude --print "Explain the main function in src/index.ts"
echo "Fix the typo in README.md" | claude --print
```

–dangerously-skip-permissions

Disables the permission system entirely. Claude Code executes all operations without asking for approval. Intended exclusively for CI/CD pipelines and fully sandboxed environments. See Chapter 21 for the security implications.

```
claude --dangerously-skip-permissions --print "Run the test suite and fix any fa
ilures"
```

–verbose

Enables verbose output, showing additional detail about Claude Code's operations including tool calls, file reads, and internal reasoning steps. Useful for debugging unexpected behavior.

```
claude --verbose
```

–output-format

Specifies the output format for non-interactive mode. Options include `text` (default), `json`, and `stream-json`.

```
claude --print --output-format json "List all exported functions in src/utils.ts
"
```

–max-turns

Limits the number of agentic turns Claude Code can take in a single interaction. Useful for controlling costs and preventing runaway automated sessions.

```
claude --max-turns 10 --print "Refactor the utils module"
```

–system-prompt

Provides a custom system prompt that overrides Claude Code's default behavior. Advanced usage for specialized workflows.

```
claude --system-prompt "You are a security auditor. Only report potential vulner
abilities."
```

–append-system-prompt

Adds to Claude Code's default system prompt rather than replacing it.

```
claude --append-system-prompt "Always suggest test cases for any code you write.
"
```

–allowedTools

Restricts which tools Claude Code can use during the session. Accepts a comma-separated list of tool names.

–disallowedTools

Prevents Claude Code from using specific tools during the session.

–resume, -r

Resumes a previous Claude Code session by its session ID. Claude Code displays recent sessions when you use this flag without specifying an ID.

```
claude --resume
claude --resume abc123
```

–continue, -c

Continues the most recent Claude Code session in the current directory.

```
claude --continue
```

Settings Configuration

Claude Code's settings are stored in JSON format at multiple levels.

Settings File Locations

- **User settings:** `~/.claude/settings.json` — applies to all your Claude Code sessions.
- **Project settings:** `.claude/settings.json` in your project root — applies to all sessions in this project, shared via version control.

- **Local project settings:** `.claude/settings.local.json` — project-specific settings not shared with the team (add to `.gitignore`).

Key Settings

```
{
  "model": "sonnet",
  "permissions": {
    "allow": [
      "Read",
      "Write",
      "Bash(npm run test)",
      "Bash(npm run build)"
    ],
    "deny": [
      "Bash(rm -rf *)",
      "Bash(sudo *)"
    ]
  },
  "env": {
    "CLAUDE_CODE_MAX_OUTPUT_TOKENS": "16384"
  }
}
```

Permission Patterns in Settings

The `permissions` object supports pattern matching for pre-approving or denying operations:

- `"Read"` — Allow all file reads.
- `"Write(src/**)"` — Allow writes to files in the src directory.
- `"Bash(npm run *)"` — Allow any npm run command.
- `"Bash(git *)"` — Allow any git command.
- `"mcp__servername__toolname"` — Allow a specific MCP tool.

MCP Server Configuration

MCP servers are configured in your settings file:

```
{
  "mcpServers": {
    "filesystem": {
```

```
      "command": "npx",
      "args": ["-y", "@modelcontextprotocol/server-filesystem", "/path/to/dir"],
      "env": {}
    },
    "github": {
      "command": "npx",
      "args": ["-y", "@modelcontextprotocol/server-github"],
      "env": {
        "GITHUB_TOKEN": "your-token-here"
      }
    }
  }
}
```

Environment Variables

These environment variables affect Claude Code's behavior when set in your shell.

ANTHROPIC_API_KEY

Your Anthropic API key for authentication. Required when not using browser-based authentication.

```
export ANTHROPIC_API_KEY="sk-ant-..."
```

CLAUDE_CODE_MAX_OUTPUT_TOKENS

Controls the maximum number of output tokens per response. Higher values allow longer responses but cost more.

CLAUDE_CODE_DISABLE_NONESSENTIAL_TRAFFIC

When set to 1, disables telemetry and non-essential network requests.

CLAUDE_CODE_USE_BEDROCK

Set to 1 to route requests through Amazon Bedrock instead of the direct Anthropic API.

CLAUDE_CODE_USE_VERTEX

Set to 1 to route requests through Google Vertex AI.

ANTHROPIC_BASE_URL

Overrides the default API endpoint. Used for enterprise deployments with custom API routing or proxy configurations.

HTTP_PROXY / HTTPS_PROXY

Standard proxy environment variables. Claude Code respects these for routing API traffic through corporate proxies.

CLAUDE_CODE_SKIP_CLAUDE_MD

When set to `1`, skips loading CLAUDE.md files. Useful for debugging configuration issues.

For CLAUDE.md templates tailored to specific project types, see Appendix B.

Appendix B: CLAUDE.md Templates

These templates provide starting points for CLAUDE.md files tailored to common project types. Copy the relevant template, customize it for your project, and iterate as your project evolves. Each template is deliberately concise — remember that your CLAUDE.md is loaded on every interaction and its token cost compounds over a session.

Web Application (React + Node.js)

```
## Project: [App Name]

### Overview
[One-sentence description]. Built with React 19 (TypeScript) frontend and Node.j
s/Express backend. PostgreSQL database with Prisma ORM.

### Architecture
- `client/` -- React SPA (Vite build)
- `server/` -- Express API server
- `shared/` -- Shared TypeScript types and validation schemas
- `prisma/` -- Database schema and migrations

### Key Conventions
- Functional components only. No class components.
- Props types named `[Component]Props`, e.g., `ButtonProps`.
- Custom hooks in `client/src/hooks/`, prefixed with `use`.
- API routes follow REST conventions: `GET /api/users`, `POST /api/users`, etc.
- All API responses use the shape `{ data: T } | { error: string, code: string }
`.
- Zod schemas in `shared/schemas/` for request validation.
- Server-side errors use `AppError` class from `server/src/errors.ts`.

### Common Tasks
- Dev server: `npm run dev` (starts both client and server)
- Tests: `npm test` (Jest for server, Vitest for client)
- Lint: `npm run lint`
- DB migration: `npx prisma migrate dev`
- DB seed: `npx prisma db seed`
- Build: `npm run build`

### Testing
```

```
- Server tests in `server/src/__tests__/`, colocated with source.
- Client tests in `client/src/**/__tests__/`.
- Use `msw` for API mocking in client tests.
- Integration tests in `tests/integration/`.

### Important Notes
- Never modify `prisma/migrations/` manually. Use `prisma migrate dev` to create
 new migrations.
- Environment variables in `.env` (not committed). See `.env.example` for requir
ed vars.
- Auth uses JWT with refresh tokens. Token logic in `server/src/auth/`.
```

Mobile Application (React Native)

```
## Project: [App Name]

### Overview
[One-sentence description]. React Native (Expo managed workflow) with TypeScript.

### Architecture
- `src/screens/` -- Screen components (one per file, named `[Name]Screen.tsx`)
- `src/components/` -- Reusable UI components
- `src/navigation/` -- React Navigation setup and type definitions
- `src/hooks/` -- Custom hooks
- `src/services/` -- API clients and external service integrations
- `src/store/` -- Zustand stores
- `src/utils/` -- Utility functions
- `src/types/` -- Shared TypeScript types

### Key Conventions
- Use Expo SDK APIs where available before reaching for native modules.
- Navigation types defined in `src/navigation/types.ts`. All screens are typed.
- API calls go through `src/services/api.ts` which handles auth headers and erro
r formatting.
- Styles use StyleSheet.create, colocated at bottom of component files.
- No inline styles except for dynamic values.
- Use `react-native-reanimated` for animations, not Animated API.

### Common Tasks
- Start dev: `npx expo start`
- Run iOS: `npx expo run:ios`
- Run Android: `npx expo run:android`
- Tests: `npm test` (Jest with React Native Testing Library)
- Lint: `npm run lint`
```

- Type check: `npx tsc --noEmit`

Testing
- Unit tests colocated as `__tests__/[Component].test.tsx`.
- Screen tests should test user flows, not implementation details.
- Mock navigation with `@react-navigation/native` mock in `jest.setup.ts`.

Important Notes
- Target iOS 16+ and Android API 28+.
- Images in `assets/images/`, optimized before committing (max 2x resolution).
- Deep links configured in `app.json` under `expo.scheme`.

Mobile Application (Flutter)

Project: [App Name]

Overview
[One-sentence description]. Flutter app targeting iOS and Android.

Architecture
- `lib/` -- Dart source code
  - `lib/screens/` -- Screen widgets
  - `lib/widgets/` -- Reusable widgets
  - `lib/models/` -- Data models (freezed + json_serializable)
  - `lib/providers/` -- Riverpod providers
  - `lib/services/` -- API and platform service classes
  - `lib/utils/` -- Utility functions and extensions
  - `lib/router/` -- GoRouter configuration

Key Conventions
- State management: Riverpod. All providers in `lib/providers/`.
- Navigation: GoRouter. Routes defined in `lib/router/app_router.dart`.
- Data classes use `freezed` with `json_serializable`. Run `dart run build_runner build` after modifying models.
- Theme defined in `lib/theme/`. Use `Theme.of(context)` for colors and text styles.
- Extensions on BuildContext in `lib/utils/context_extensions.dart` for common operations.

Common Tasks
- Run app: `flutter run`
- Tests: `flutter test`
- Build runner: `dart run build_runner build --delete-conflicting-outputs`
- Lint: `flutter analyze`

```
- Build APK: `flutter build apk --release`
- Build iOS: `flutter build ios --release`

### Testing
- Widget tests in `test/` mirroring `lib/` structure.
- Use `ProviderScope` overrides for mocking providers in tests.
- Golden tests for critical UI components in `test/goldens/`.

### Important Notes
- Minimum Flutter SDK: 3.24. Check with `flutter --version`.
- API base URL configured via `--dart-define` flags, not hardcoded.
- Localization files in `lib/l10n/`. Run `flutter gen-l10n` after adding translations.
```

CLI Tool (Python)

```
## Project: [Tool Name]

### Overview
[One-sentence description]. Python CLI tool using Click for argument parsing.

### Architecture
- `src/[package_name]/` -- Main package
  - `cli.py` -- Click command definitions (entry point)
  - `core.py` -- Core business logic
  - `config.py` -- Configuration loading and validation
  - `utils.py` -- Utility functions
  - `models.py` -- Pydantic data models
- `tests/` -- Test directory mirroring src structure

### Key Conventions
- Python 3.11+. Type hints on all function signatures.
- Use `pathlib.Path` for file paths, never `os.path`.
- CLI commands use Click decorators. Group commands logically.
- Configuration loaded from `~/.config/[tool]/config.toml` with env var overrides.
- Errors displayed to user via `click.echo` with styled output. Internal errors use logging.
- All data structures are Pydantic BaseModel subclasses.

### Common Tasks
- Install dev: `pip install -e ".[dev]"`
- Tests: `pytest`
- Tests with coverage: `pytest --cov=src`
```

```markdown
- Lint: `ruff check .`
- Format: `ruff format .`
- Type check: `mypy src/`
- Build: `python -m build`

### Testing
- Tests use pytest with fixtures in `tests/conftest.py`.
- CLI tests use Click's `CliRunner`.
- Use `tmp_path` fixture for file system tests.
- Mock external calls with `responses` or `unittest.mock`.

### Important Notes
- Package managed with `pyproject.toml`. No `setup.py`.
- Dependencies pinned in `requirements-lock.txt`. Dev deps in `requirements-dev.txt`.
- Entry point defined in `pyproject.toml` under `[project.scripts]`.
```

CLI Tool (Go)

```markdown
## Project: [Tool Name]

### Overview
[One-sentence description]. Go CLI tool using Cobra for command structure.

### Architecture
- `cmd/` -- Cobra command definitions
  - `cmd/root.go` -- Root command and global flags
  - `cmd/[subcommand].go` -- One file per subcommand
- `internal/` -- Internal packages (not importable by other modules)
  - `internal/config/` -- Configuration handling
  - `internal/[domain]/` -- Domain logic packages
- `pkg/` -- Public packages (if any are intended for library use)
- `main.go` -- Entry point, calls `cmd.Execute()`

### Key Conventions
- Go 1.22+. Use standard library where possible before adding dependencies.
- Error handling: return errors, do not panic. Wrap with `fmt.Errorf("context: %w", err)`.
- Cobra commands follow the pattern: `[tool] [verb] [noun] --flags`.
- Configuration via Viper: config file, env vars, flags (in that priority order).
- Structured logging with `slog`.
- Context propagation: pass `context.Context` as first parameter.

### Common Tasks
```

```
- Build: `go build -o bin/[tool] .`
- Test: `go test ./...`
- Test verbose: `go test -v ./...`
- Lint: `golangci-lint run`
- Run: `go run . [command]`

### Testing
- Table-driven tests preferred.
- Test files colocated: `foo_test.go` next to `foo.go`.
- Use `testdata/` directories for fixture files.
- Mock interfaces, not concrete types.

### Important Notes
- Module path: `github.com/[org]/[tool]`.
- Release builds use goreleaser. Config in `.goreleaser.yaml`.
- Do not vendor dependencies. Use Go module proxy.
```

CLI Tool (Rust)

```
## Project: [Tool Name]

### Overview
[One-sentence description]. Rust CLI tool using clap for argument parsing.

### Architecture
- `src/main.rs` -- Entry point, CLI arg parsing
- `src/cli.rs` -- Clap derive structs for CLI definition
- `src/commands/` -- One module per subcommand
- `src/lib.rs` -- Library root (core logic)
- `src/config.rs` -- Configuration handling
- `src/error.rs` -- Custom error types using thiserror

### Key Conventions
- Rust edition 2024. MSRV: 1.85.
- Error handling: custom error enum with `thiserror`. Use `anyhow` in `main`, `thiserror` in library code.
- Clap derive API for CLI definition. Subcommands as enum variants.
- Use `serde` for serialization. Config files in TOML format.
- Logging with `tracing`. Initialize subscriber in `main`.

### Common Tasks
- Build: `cargo build`
- Test: `cargo test`
- Run: `cargo run -- [args]`
```

- Lint: `cargo clippy -- -D warnings`
- Format: `cargo fmt`
- Release build: `cargo build --release`

Testing
- Unit tests in `#[cfg(test)]` modules at bottom of source files.
- Integration tests in `tests/`.
- Use `assert_cmd` and `predicates` for CLI integration tests.
- Use `tempfile` for filesystem tests.

Important Notes
- CI runs `cargo clippy -- -D warnings` and `cargo fmt -- --check`. Fix before pushing.
- Dependencies: prefer well-maintained crates with minimal transitive deps.
- Binary size matters. Use `cargo bloat` to check. Strip in release profile.

Library / SDK

Project: [Library Name]

Overview
[One-sentence description]. Published to [npm/PyPI/crates.io/etc.] as `[package-name]`.

Architecture
- `src/` -- Source code
  - `src/index.ts` -- Public API exports (barrel file)
  - `src/client.ts` -- Main client class
  - `src/types.ts` -- Public TypeScript types
  - `src/internal/` -- Internal implementation (not exported)
  - `src/errors.ts` -- Custom error classes

Key Conventions
- Every public API must have JSDoc comments with `@example` tags.
- Semantic versioning. Breaking changes require major version bump.
- All public types exported from `src/index.ts`.
- No runtime dependencies if possible. Peer dependencies where required.
- Async methods return Promises, never callbacks.
- Error classes extend `[LibName]Error` base class.
- Configuration via constructor options object with sensible defaults.

Common Tasks
- Build: `npm run build` (outputs to `dist/`)
- Test: `npm test`

```markdown
- Lint: `npm run lint`
- Docs: `npm run docs` (generates from JSDoc)
- Publish: `npm publish` (CI only)

### Testing
- Unit tests colocated as `[file].test.ts`.
- Integration tests in `tests/integration/`.
- Test public API surface, not internal implementation.
- Use `nock` for HTTP mocking. No real network calls in tests.
- 90%+ coverage target on public API.

### Important Notes
- This is a library. Do not add dependencies lightly -- every dependency is inherited by consumers.
- Support Node.js 20+ and modern browsers (last 2 versions).
- Bundle uses tsup. Config in `tsup.config.ts`.
- Changesets for versioning: run `npx changeset` when making changes.
```

Monorepo

```markdown
## Project: [Project Name]

### Overview
[One-sentence description]. Monorepo managed with [Turborepo/Nx/pnpm workspaces].

### Architecture
- `apps/` -- Deployable applications
  - `apps/web/` -- Next.js web application
  - `apps/api/` -- Express API server
  - `apps/mobile/` -- React Native app
- `packages/` -- Shared packages
  - `packages/ui/` -- Shared UI component library
  - `packages/shared/` -- Shared types and utilities
  - `packages/config/` -- Shared ESLint, TypeScript, and other configs
- `tooling/` -- Build and development tooling

### Key Conventions
- Package manager: pnpm. Always use `pnpm` commands, never `npm` or `yarn`.
- Each package has its own `package.json`, `tsconfig.json`, and test setup.
- Cross-package imports use workspace protocol: `"@[scope]/[package]": "workspace:*"`.
- Shared types go in `packages/shared`. App-specific types stay in the app.
- UI components in `packages/ui` must be app-agnostic.
```

```
### Common Tasks
- Install: `pnpm install`
- Dev all: `pnpm dev` (runs all apps in dev mode)
- Dev specific: `pnpm --filter web dev`
- Build all: `pnpm build`
- Build specific: `pnpm --filter api build`
- Test all: `pnpm test`
- Test specific: `pnpm --filter @[scope]/ui test`
- Lint: `pnpm lint`
- Add dep to package: `pnpm --filter [package] add [dep]`

### Testing
- Each package owns its tests. Run from package root or via workspace filter.
- Shared test utilities in `packages/config/test-utils`.
- CI runs `pnpm test` at root, which delegates to all packages.

### Important Notes
- Turbo caching is enabled. If tests seem stale, run `pnpm turbo clean`.
- New packages need to be added to `pnpm-workspace.yaml`.
- Do not create circular dependencies between packages. The dependency graph must be a DAG.
- Environment variables: each app has its own `.env`. Shared env vars in root `.env`.
```

Data Pipeline

```
## Project: [Pipeline Name]

### Overview
[One-sentence description]. Python data pipeline using [Airflow/Dagster/Prefect] for orchestration.

### Architecture
- `dags/` or `pipelines/` -- Pipeline/DAG definitions
- `src/[package]/`
  - `extractors/` -- Data extraction from sources
  - `transformers/` -- Data transformation logic
  - `loaders/` -- Data loading to destinations
  - `models/` -- Pydantic models for data schemas
  - `utils/` -- Shared utilities (logging, config, connections)
- `sql/` -- SQL templates and migrations
- `tests/` -- Test directory
- `config/` -- Pipeline configuration files (YAML)
```

```
### Key Conventions
- Python 3.11+. Type hints required.
- Each pipeline step is a pure function: input data in, output data out. No side
 effects in transformation logic.
- Database connections managed via connection pool in `src/[package]/utils/db.py
`.
- SQL queries in `sql/` directory, loaded as templates. No SQL strings in Python
 code.
- Configuration per environment in `config/[env].yaml`.
- Logging: structured JSON via `structlog`. Include pipeline_name and step_name
in context.
- Data validation with Pydantic at extraction and before loading.

### Common Tasks
- Run pipeline locally: `python -m [package].cli run [pipeline_name]`
- Test: `pytest`
- Lint: `ruff check .`
- Format: `ruff format .`
- Type check: `mypy src/`
- DB migration: `alembic upgrade head`

### Testing
- Unit tests for transformers (pure functions, easy to test).
- Integration tests for extractors and loaders with test database.
- Use `pytest-docker` for test database setup.
- Fixtures in `tests/fixtures/` with sample data files.

### Important Notes
- Never commit connection strings, API keys, or credentials. Use environment var
iables.
- Pipeline configs in `config/` are environment-specific. Do not hardcode enviro
nment assumptions.
- Large data processing uses chunked/streaming patterns. Do not load entire data
sets into memory.
- Idempotency required: every pipeline step must be safe to re-run.
```

For a directory of MCP servers that extend Claude Code's capabilities, see Appendix C.

Appendix C: MCP Server Directory

The Model Context Protocol (MCP) extends Claude Code's capabilities by connecting it to external tools, services, and data sources. This directory covers the most useful MCP servers as of early 2026, organized by category. Each entry includes a description, typical use cases, and basic setup guidance.

MCP servers are configured in your Claude Code settings file. See Appendix A for the configuration format.

Official MCP Servers

These servers are maintained by Anthropic or are part of the official MCP specification repository.

Filesystem Server

Package: `@modelcontextprotocol/server-filesystem`

Gives Claude Code structured file system access beyond its built-in capabilities. Useful when you want access to directories outside the current project, or when working with MCP-based workflows that need explicit file system tools.

```
{
  "command": "npx",
  "args": ["-y", "@modelcontextprotocol/server-filesystem", "/path/to/allowed/di
rectory"]
}
```

Use cases: Cross-project file access, working with configuration files in system directories, managing assets in external locations.

GitHub Server

Package: `@modelcontextprotocol/server-github`

Connects Claude Code to the GitHub API. Creates issues, manages pull requests, searches repositories, reads file contents from remote repos, and interacts with GitHub Actions.

```
{
  "command": "npx",
  "args": ["-y", "@modelcontextprotocol/server-github"],
  "env": {
    "GITHUB_TOKEN": "ghp_your_token_here"
  }
}
```

Use cases: Creating PRs from Claude Code, triaging issues, searching codebases on GitHub, reviewing pull requests, managing releases.

GitLab Server

Package: `@modelcontextprotocol/server-gitlab`

Similar to the GitHub server but for GitLab instances. Supports both gitlab.com and self-hosted GitLab.

```
{
  "command": "npx",
  "args": ["-y", "@modelcontextprotocol/server-gitlab"],
  "env": {
    "GITLAB_TOKEN": "glpat-your_token_here",
    "GITLAB_URL": "https://gitlab.example.com"
  }
}
```

Use cases: Same as GitHub server, for teams using GitLab.

PostgreSQL Server

Package: `@modelcontextprotocol/server-postgres`

Provides read-only access to PostgreSQL databases. Claude Code can query your schema and data to answer questions, generate queries, and debug data issues.

```
{
  "command": "npx",
  "args": ["-y", "@modelcontextprotocol/server-postgres", "postgresql://user:pas
s@localhost:5432/dbname"]
}
```

Use cases: Exploring database schemas, debugging data issues, generating SQL queries, understanding data relationships.

Security note: Use read-only database credentials. Never connect Claude Code to a production database with write access.

SQLite Server

Package: `@modelcontextprotocol/server-sqlite`

Read and write access to SQLite databases. Useful for local development databases and application data stores.

```
{
  "command": "npx",
  "args": ["-y", "@modelcontextprotocol/server-sqlite", "/path/to/database.db"]
}
```

Memory Server

Package: `@modelcontextprotocol/server-memory`

Provides persistent key-value storage Claude Code can read and write across sessions. Acts as a simple knowledge base that persists beyond conversation history.

```
{
  "command": "npx",
  "args": ["-y", "@modelcontextprotocol/server-memory"]
}
```

Use cases: Storing project decisions, maintaining a running list of TODOs across sessions, persisting learned preferences.

Fetch Server

Package: `@modelcontextprotocol/server-fetch`

Lets Claude Code make HTTP requests and read web content. Fetches URLs and returns content in a format Claude Code can process.

```
{
  "command": "npx",
  "args": ["-y", "@modelcontextprotocol/server-fetch"]
}
```

Use cases: Reading API documentation, checking endpoint responses, fetching remote configuration, accessing web-based resources.

Puppeteer Server

Package: `@modelcontextprotocol/server-puppeteer`

Provides browser automation capabilities. Claude Code can navigate web pages, take screenshots, click elements, and fill forms.

```
{
  "command": "npx",
  "args": ["-y", "@modelcontextprotocol/server-puppeteer"]
}
```

Use cases: Testing web applications, debugging UI issues, automating browser-based workflows, taking screenshots for documentation.

Community MCP Servers

These servers are built and maintained by the community. Quality and maintenance vary — evaluate each before adding it to your workflow.

Slack Server

Connects Claude Code to Slack workspaces. Can read messages, post to channels, and search conversation history.

Use cases: Searching Slack for context on decisions, posting deployment notifications, pulling conversation context into development sessions.

Linear Server

Integrates with Linear for project management. Claude Code can read issues, create tasks, update status, and search the backlog.

Use cases: Creating issues from code TODOs, updating task status during development, pulling issue context into coding sessions.

Jira Server

Similar to the Linear server but for Jira. Supports reading and creating issues, managing sprints, and searching.

Use cases: Teams using Jira for project management who want Claude Code to interact with their issue tracker.

Sentry Server

Connects to Sentry for error monitoring. Claude Code can search errors, read stack traces, and access error context.

Use cases: Debugging production errors by pulling Sentry context directly into your Claude Code session.

Docker Server

Provides Docker management capabilities. Claude Code can list containers, read logs, manage images, and interact with Docker Compose.

Use cases: Debugging containerized applications, managing local development environments, reading container logs.

Kubernetes Server

Connects to Kubernetes clusters for read access to pods, services, deployments, and logs.

Use cases: Debugging Kubernetes deployments, reading pod logs, understanding cluster state during development.

Security note: Use read-only RBAC roles. Do not give Claude Code cluster-admin access.

AWS Server

Provides access to AWS services. Scope varies by implementation — some cover S3, others include EC2, Lambda, and more.

Use cases: Managing cloud resources during development, reading CloudWatch logs, checking deployment status.

Security note: Use IAM roles with minimal permissions. Never use root credentials.

Notion Server

Connects to Notion workspaces. Claude Code can read pages, search content, and create or update pages.

Use cases: Pulling design documents into development context, updating technical documentation, searching knowledge bases.

Brave Search Server

Provides web search capabilities through the Brave Search API.

```
{
  "command": "npx",
  "args": ["-y", "@modelcontextprotocol/server-brave-search"],
  "env": {
    "BRAVE_API_KEY": "your_api_key_here"
  }
}
```

Use cases: Searching for documentation, finding solutions to errors, researching libraries and APIs.

Exa Search Server

AI-optimized web search through the Exa API. Returns semantically relevant results rather than keyword matches.

Use cases: Finding relevant code examples, researching best practices, discovering libraries.

Choosing and Managing MCP Servers

Selection Criteria

When deciding whether to add an MCP server, consider:

- **Do you need it regularly?** If you will use it once a month, connect it on demand rather than permanently.

- **What access does it require?** Each server's credentials are a security surface. Only connect servers whose access requirements you are comfortable with.
- **Is it maintained?** Check the repository's last commit date, open issues, and community activity.
- **What is the token cost?** Some MCP servers return verbose output that consumes context tokens. Test before committing.

Security Best Practices

- Use read-only credentials wherever possible.
- Create dedicated API tokens for Claude Code rather than sharing your personal tokens.
- Rotate credentials on a regular schedule.
- Audit which MCP servers are configured across your team.
- Remove servers you are not actively using.

Performance Considerations

Each connected MCP server adds startup time to your Claude Code sessions as connections are established. Servers that maintain persistent connections (databases, message queues) consume system resources.

Keep your active MCP server list lean. A focused set of three to five servers you use daily beats a dozen that mostly sit idle but slow your session startup.

For solutions to common Claude Code problems, see Appendix D.

Appendix D: Troubleshooting

When Claude Code misbehaves, the fix is usually simple — if you know where to look. This appendix covers the most common issues, organized by category, with concrete solutions for each.

Start every troubleshooting session with `/doctor`. It catches most configuration and connectivity problems automatically.

Authentication Issues

"Invalid API key" or "Authentication failed"

Cause: Your API key is missing, expired, or malformed.

Fix: 1. Verify the key is set: check your environment with `echo $ANTHROPIC_API_KEY` (should start with `sk-ant-`). 2. If using a settings file, verify the key is correctly formatted in your configuration. 3. Try re-authenticating with `/login`. 4. Generate a new key from the Anthropic console if the current one may be compromised or expired.

"Rate limit exceeded"

Cause: You have hit your API usage tier's rate limit.

Fix: 1. Wait a few minutes and retry. Rate limits reset on a rolling window. 2. If this happens often, check your Anthropic account tier and consider upgrading. 3. For teams, make sure multiple developers are not sharing a single API key.

Authentication works but model is unavailable

Cause: Your account may not have access to the requested model tier.

Fix: 1. Switch to a different model: `/model sonnet` or `/model haiku`. 2. Check your Anthropic account dashboard for model access details. 3. Some models require specific account tiers or enterprise agreements.

Connection and Network Issues

"Connection refused" or timeout errors

Cause: Network connectivity issue between your machine and the Anthropic API.

Fix: 1. Check your internet connection with a simple test: `curl -I https://api.anthropic.com`. 2. Behind a corporate proxy, set `HTTP_PROXY` and `HTTPS_PROXY` environment variables. 3. On a VPN, try disconnecting — some VPNs block or throttle API traffic. 4. Check whether `ANTHROPIC_BASE_URL` is set to an incorrect value.

Slow responses

Cause: Network latency, large context windows, or model load.

Fix: 1. Check your context window size with `/cost`. Large contexts mean slower responses. 2. Use `/compact` to shrink context. 3. Switch to Haiku for faster responses on simple tasks. 4. If using a proxy, test without it to isolate the bottleneck.

"SSL certificate error"

Cause: Corporate proxy performing SSL inspection, or outdated CA certificates.

Fix: 1. Behind a corporate proxy with SSL inspection, configure your system's CA bundle to include the proxy's certificate. 2. Update your system's CA certificates. 3. As a last resort (not recommended for production), set `NODE_TLS_REJECT_UNAUTHORIZED=0` — understand that this disables SSL verification entirely.

Permission Issues

Claude Code keeps asking for the same permission

Cause: Permissions are session-scoped by default and reset when you start a new session.

Fix: 1. Add frequently used permissions to your settings file (see Appendix A for the permissions format). 2. Use project-level settings (`.claude/settings.json`) for project-specific permissions. 3. Example: to permanently allow test runs, add `"Bash(npm run test)"` to your `permissions.allow` array.

"Permission denied" when accessing files

Cause: Operating system file permissions, not Claude Code permissions. Claude Code runs as your user and inherits your OS-level access.

Fix: 1. Check file permissions: `ls -la [file]`. 2. Confirm the file is not locked by another process. 3. On Windows, check that the file is not marked read-only. 4. Verify you have access to the directory containing the file.

Commands fail silently after permission denial

Cause: You denied a permission and Claude Code proceeded without the operation.

Fix: 1. Read the conversation output carefully — Claude Code should note when an operation was skipped due to a denied permission. 2. Re-run the task and approve the required permission. 3. Use `--verbose` to see detailed operation logs.

CLAUDE.md Issues

CLAUDE.md not loading

Cause: File is in the wrong location, incorrectly named, or loading is disabled.

Fix: 1. Verify file name is exactly `CLAUDE.md` (case-sensitive on Linux/macOS). 2. Verify location: project root for project settings, `~/.claude/` for user-level settings. 3. Check that `CLAUDE_CODE_SKIP_CLAUDE_MD` environment variable is not set to `1`. 4. Run `/status` to see which CLAUDE.md files are loaded.

CLAUDE.md instructions being ignored

Cause: Instructions are too vague, contradictory, or buried in too much text.

Fix: 1. Keep instructions specific and actionable. "Use camelCase" works; "follow best practices" does not. 2. Put the most important instructions at the top of the file. 3. Remove outdated or contradictory instructions. 4. If the file is long, split it — critical instructions in CLAUDE.md, reference material in separate files that Claude Code reads on demand.

CLAUDE.md making sessions expensive

Cause: Verbose CLAUDE.md files add tokens to every interaction.

Fix: 1. Audit your CLAUDE.md for unnecessary content. Every word costs tokens on every interaction. 2. Move reference material (long code examples, detailed API documentation) to separate files. 3. Aim for under 500 words. If you need more, ask whether every section is truly needed on every interaction.

MCP Server Issues

MCP server fails to connect

Cause: Server package not installed, incorrect configuration, or missing credentials.

Fix: 1. Verify the server package is available: `npx -y [package-name] --help`. 2. Check your settings file for typos in the server configuration. 3. Verify environment variables (API keys, tokens) are set correctly in the server's `env` configuration. 4. Run `/mcp` to check server connection status. 5. Run the server command manually in your terminal to see error output.

MCP server connected but tools not appearing

Cause: Server started but failed to register its tools, or tool names do not match expectations.

Fix: 1. Run `/mcp` to see which tools are registered. 2. Restart Claude Code — some MCP servers need a fresh connection. 3. Check the server's documentation for correct tool names. 4. Update the server package to the latest version.

MCP server causing slow startup

Cause: Server takes a long time to initialize, or makes network calls during startup.

Fix: 1. Remove MCP servers you are not actively using. 2. Check if the server makes network calls during initialization (database connections, API verification). 3. Consider starting Claude Code without the slow server and adding it mid-session if needed.

IDE Integration Issues

VS Code extension not connecting to Claude Code

Cause: Extension version mismatch, Claude Code not running, or port conflict.

Fix: 1. Update both the VS Code extension and Claude Code to their latest versions. 2. Restart VS Code after updating. 3. Check the VS Code output panel (View > Output, select Claude Code from the dropdown) for error messages. 4. Disable other AI coding extensions that might conflict.

Editor commands not reflecting in Claude Code

Cause: Communication between the IDE extension and Claude Code is broken.

Fix: 1. Restart the Claude Code session from within the IDE. 2. Check that the IDE extension points to the correct Claude Code installation. 3. Confirm no firewall or security software is blocking local communication.

Terminal integration issues

Cause: Shell profile not configured for Claude Code integration.

Fix: 1. Run `/terminal-setup` to configure your shell. 2. Restart your terminal after setup. 3. Verify your shell profile (`.bashrc`, `.zshrc`, etc.) includes the Claude Code integration lines.

Performance Issues

Claude Code using excessive memory

Cause: Large context windows, many MCP server connections, or long-running sessions.

Fix: 1. Start a new session to reset memory usage. 2. Cut the number of connected MCP servers. 3. Use `/compact` regularly in long sessions. 4. Close other memory-intensive applications if your system is constrained.

Responses getting worse over long sessions

Cause: Context window approaching its limit, causing older context to be truncated or poorly summarized.

Fix: 1. Start a fresh session. The most reliable fix. 2. Use `/compact` to create a clean summary of relevant context. 3. For complex tasks, break work into multiple sessions with clear handoff notes.

File operations are slow

Cause: Very large files, many files in the project directory, or slow disk.

Fix: 1. Use `.claudeignore` to exclude large directories (e.g., `node_modules`, `.git`, build output). 2. Make sure `.gitignore` is properly configured — Claude Code uses it to filter file indexing. 3. Work in the most specific directory possible rather than a parent directory with many subdirectories.

Common Error Messages

"Context window exceeded"

Your conversation plus file context exceeds the model's context limit. Use `/compact`, start a new session, or reduce the number of files in context.

"Output truncated"

The response hit the output token limit. Ask Claude Code to continue, or increase `CLAUDE_CODE_MAX_OUTPUT_TOKENS`.

"Tool execution failed"

A shell command or MCP tool call failed. Check the error output for specifics. Common causes: missing dependencies, wrong paths, or insufficient permissions.

"Model overloaded"

The model is experiencing high demand. Wait a moment and retry. If persistent, try a different model.

When All Else Fails

If none of the above work:

1. **Update Claude Code** to the latest version.
2. **Check the Anthropic status page** for service outages.
3. **Reproduce with `--verbose`** to capture detailed logs.
4. **Start clean:** new terminal, new session, minimal configuration. If it works clean, add your customizations back one at a time to find the culprit.
5. **Search the community** — forums, GitHub issues, and Discord often have solutions for edge cases.
6. **Report the issue** to Anthropic with reproduction steps and verbose logs.

For definitions of terms used throughout this book, see Appendix E.

Appendix E: Glossary

Terms are defined in the context of Claude Code and AI-assisted development. Where a term has a broader meaning in computer science or AI, the definition here focuses on its practical relevance to Claude Code users.

Agentic coding. A mode of AI-assisted development where the AI tool takes autonomous actions — reading files, executing commands, making edits — rather than simply generating text. Claude Code operates as an agentic coding tool, performing real operations on your system rather than suggesting code for you to copy-paste.

API key. A secret string that authenticates your requests to the Anthropic API. Claude Code uses your API key to access Claude models. Treat it like a password — do not share it, commit it to version control, or expose it in logs.

Bedrock. Amazon Bedrock, a managed service for running foundation models on AWS. Claude Code can be configured to route requests through Bedrock using the `CLAUDE_CODE_USE_BEDROCK` environment variable, which is common in enterprise environments with AWS-centric infrastructure.

CLAUDE.md. A Markdown file that provides persistent instructions and context to Claude Code. Loaded automatically at the start of every session. Can exist at the user level (`~/.claude/CLAUDE.md`), project level (project root), or nested within subdirectories. See Chapter 8 and Appendix B.

.claudeignore. A file using `.gitignore` syntax that tells Claude Code to ignore specific files or directories. Prevents sensitive or irrelevant files from being included in Claude Code's context.

Claude Max. A subscription plan from Anthropic ($100–$200/month) that provides unlimited Claude Code usage without per-token billing. An alternative to pay-per-token API access for heavy users.

Compact. The `/compact` slash command that summarizes conversation history to reduce context window size. Essential for managing costs and maintaining response quality in long sessions.

Context window. The total amount of text (measured in tokens) that a model can process in a single interaction. Includes the system prompt, conversation history, file contents, tool results, and the current prompt. Larger context windows enable more complex tasks but cost more and can slow responses.

Conversation history. The accumulated record of prompts and responses in a Claude Code session. Sent as input with each new interaction, which means the cost of each interaction grows as the session progresses.

Dangerously skip permissions. The `--dangerously-skip-permissions` CLI flag that disables Claude Code's permission system. Intended for CI/CD and fully sandboxed environments only. See Chapter 21.

Diff. A representation of changes between two versions of a file or set of files. Claude Code presents diffs when it modifies files, and reviewing diffs is a critical safety practice before committing AI-generated changes.

Embedding. A numerical representation of text that captures its semantic meaning. Used internally by search and retrieval systems. Not directly exposed in Claude Code but relevant to understanding how some MCP servers and RAG systems work.

Haiku 4.5. The smallest and cheapest model in the current Claude lineup. Fastest response times, best for simple tasks like formatting, boilerplate generation, and quick questions.

Headless mode. Running Claude Code without an interactive terminal, typically in CI/CD pipelines or automated scripts. Uses `--print` flag and `--dangerously-skip-permissions`. See Chapter 19.

Idempotent. An operation that produces the same result whether executed once or multiple times. Important when Claude Code re-runs commands or when using Claude Code in automated pipelines where retries may occur.

Input tokens. Tokens sent to the model as part of a request. Includes your prompt, conversation history, file contents, CLAUDE.md, system prompts, and tool results. Billed at a lower rate than output tokens.

MCP (Model Context Protocol). An open protocol that allows AI tools like Claude Code to connect to external services and data sources. MCP servers expose tools that Claude Code can call, extending its capabilities beyond file system access and shell commands. See Chapters 14–15 and Appendix C.

MCP server. A program that implements the Model Context Protocol and exposes tools to Claude Code. Servers can provide access to databases, APIs, cloud services, and other external systems. Configured in Claude Code's settings files.

MCP tool. A specific capability exposed by an MCP server. For example, a GitHub MCP server might expose tools for creating issues, listing pull requests, and searching code.

Model. A large language model that processes text input and generates text output. Claude Code supports multiple models: Opus 4.7 (most capable), Sonnet 4.6 (balanced), and Haiku 4.5 (cheapest). The model is the engine that powers Claude Code's intelligence.

Opus 4.7. The most capable and most expensive model in the Claude family. Best for complex reasoning, architectural decisions, and tasks requiring the highest quality output.

Output tokens. Tokens generated by the model as a response. Includes code, explanations, and tool call requests. Billed at a higher rate than input tokens.

Permission. An authorization granted to Claude Code to perform a specific operation: reading a file, writing a file, executing a shell command, or calling an MCP tool. Permissions can be granted per-interaction, per-session, or permanently via settings.

Print mode. Non-interactive mode activated with the `--print` or `-p` flag. Claude Code processes a single prompt and outputs the response, then exits. Useful for scripting and automation.

Prompt. The text input you provide to Claude Code. In the context of Claude Code, this includes both the visible text you type and the invisible context (CLAUDE.md, conversation history, system prompts) sent alongside it.

Prompt engineering. The practice of crafting prompts to elicit better responses from AI models. In Claude Code, this means writing clear, specific instructions that include relevant context and constraints.

RAG (Retrieval-Augmented Generation). A technique that augments AI responses by first retrieving relevant documents from a knowledge base. Some MCP servers implement RAG patterns to give Claude Code access to large document collections without loading everything into the context window.

Sandbox. An isolated environment that restricts what a program can access. Claude Code does not run in a sandbox by default — it has full access to your user account's permissions. Sandboxing is your responsibility when running Claude Code in sensitive environments.

Session. A single Claude Code conversation from start to finish. Sessions maintain conversation history, permission grants, and loaded context. Starting a new session resets all of these.

Settings file. A JSON configuration file that controls Claude Code's behavior. Exists at user level (`~/.claude/settings.json`), project level (`.claude/settings.json`), and local project level (`.claude/settings.local.json`). See Appendix A.

Slash command. A command entered during a Claude Code session by typing `/` followed by the command name. Examples: `/cost`, `/compact`, `/help`, `/model`. See Appendix A for the full list.

Sonnet 4.6. The balanced model in the Claude family. Offers strong capabilities at a moderate price point. Recommended as the default model for most development work.

System prompt. Instructions that define Claude Code's behavior, loaded before your conversation begins. Includes Claude Code's built-in instructions plus your CLAUDE.md content. Not directly visible but influences every response.

Token. The fundamental unit of text processing for language models. Roughly four characters of English text or three-quarters of a word. Code tokenizes differently than prose. All Claude Code costs are measured in tokens.

Tool. A capability that Claude Code can invoke to interact with the outside world. Built-in tools include file reading, file writing, shell command execution, and web search. MCP servers add additional tools.

Tool call. A specific invocation of a tool by Claude Code. When Claude Code reads a file, runs a command, or uses an MCP tool, each action is a tool call. Tool calls and their results consume tokens.

Verbose mode. Activated with the `--verbose` flag. Shows detailed information about Claude Code's operations including tool calls, file reads, and internal processing steps. Essential for debugging.

Vertex AI. Google Cloud's AI platform. Claude Code can be configured to route requests through Vertex AI using the `CLAUDE_CODE_USE_VERTEX` environment variable. Common in enterprises with Google Cloud infrastructure.

This glossary covers the essential terminology. For broader AI and machine learning concepts, see "Claude: The Definitive Guide" from Finnoybu Press (press.finnoybu.org).

www.ingramcontent.com/pod-product-compliance
Ingram Content Group UK Ltd.
Pitfield, Milton Keynes, MK11 3LW, UK
UKHW051117290926
15471UKWH00010B/1720

9 798950 533037